Geoff Engelstein is an adjunct professor of Board Game Design at the NYU Game Center. He has spoken at a variety of venues, including Pax, GDC, Gencon, Rutgers and USC. He has degrees in Physics and Electrical Engineering from the Massachusetts Institute of Technology, and is currently the president of Mars International, a design engineering firm.

Since 2007 he has been a contributor to *The Dice Tower*, the leading table-top game podcast, with 'GameTek' – a series on the math, science and psychology of games. Since 2011 he has hosted *Ludology*, a weekly podcast. He is also an award-winning table-top game designer, whose games include *The Ares Project*, *Space Cadets*, *The Fog of War* and *Survive: Space Attack*, many of which are co-designed with his children, Brian and Sydney.

GAMETEK

Geoffrey Engelstein

HarperCollins*Publishers*

This is a book about board games and the benefits of board games. The games that are referred to in this book are included for referencing purposes only. By including references to games in this book neither HarperCollins nor the author have sought to imply that they have the endorsement, sponsorship or approval of the relevant rightsholders.

To Susan: Alea iacta est!

HarperCollins*Publishers*

First published in Australia in 2019
by HarperCollins*Publishers* Australia Pty Limited
ABN 36 009 913 517
harpercollins.com.au

Copyright © Mind Bullet Games LLC 2019

The right of Geoffrey Engelstein to be identified as the author of this work has been asserted by him in accordance with the *Copyright Amendment (Moral Rights) Act 2000.*

HarperCollins*Publishers*
Level 13, 201 Elizabeth Street, Sydney NSW 2000, Australia
Unit D1, 63 Apollo Drive, Rosedale, Auckland 0632, New Zealand
A 53, Sector 57, Noida, UP, India
1 London Bridge Street, London, SE1 9GF, United Kingdom
Bay Adelaide Centre, East Tower, 22 Adelaide Street West, 41st floor, Toronto, Ontario M5H 4E3, Canada
195 Broadway, New York NY 10007, USA

A catalogue record for this book is available
from the National Library of Australia

ISBN: 978 1 4607 5737 6 (paperback)
ISBN: 978 1 4607 1112 5 (ebook)

Cover design by Hazel Lam, HarperCollins Design Studio
Cover images by shutterstock.com
Author photo by David Stickles
Playing cards on pages 177 and 178 are from Shutterstock
Typeset in Perpetua by Kelli Lonergan

Printed and bound by CPI Group (UK) Ltd, Croydon, CR0 4YY

Contents

Introduction

I've been obsessed with science and math and games my entire life, and I voraciously read many, many books about them when I was growing up.

Two seminal works for me are *Gödel, Escher Bach: An Eternal Golden Braid* by Douglas Hofstadter, and *The Ascent of Man* by Jacob Bronowski. I encountered both of these when I was 14, and I was blown away by the way that they each took disparate concepts and wove them together into a coherent picture of the universe and our place in it. Part detective stories, part towering works of construction, these books show how everything is connected to everything else.

When Tom Vasel, one of the hosts of the premier podcast about board games in the US, *The Dice Tower*, announced that they were looking for new segment ideas, I thought it would be a fun project to try to come up with one. The opportunity to try to do what Hofstadter and Bronowski had done so brilliantly, in my own small way, was irresistible, and almost immediately I came up with the idea of using games to explain math, science and psychology. But I was very concerned that I wouldn't be able to keep the segment going, either for the short or long term. My life is strewn with the carcasses of projects begun and abandoned. And

I was fairly confident that I could only get, at most, twenty segments out of the idea.

To convince myself that it was viable, I decided to write and record three segments, to give myself a little buffer, and also to give Tom an idea of what I had in mind. Those first three segments were 'Dice and Luck', 'Reviews', and a discussion of the potential to use touchscreen tables for playing games. The first two have been included in this book (see Chapters 9 and 7). The third, my discussion of the then-prototype Microsoft Surface Table, is a little dated now.

The first 'GameTek' segment aired on *The Dice Tower* episode 104 in August 2007. And my estimate of 20 segments was obviously way off base. I don't have an exact count, but I am pretty sure there have been over 200 at this point. Just when I think I've run out of things to talk about, some new research or concept arrives to prove me wrong. And over ten years later, I was thrilled to have the opportunity to collect many of the best of them here in this book.

Over the years of broadcasting 'GameTek', I have come to realise that games are, in many ways, an ideal way to tie together the disparate threads of math and science. The idea of 'play', which is at the heart of games, is a fundamental human behaviour — beyond humans, actually, as any observation of young animals will readily demonstrate. It is as fundamental as eating or breathing. Yet because we associate playing with fun, to use games — both familiar and new — as a way to introduce concepts helps get past the barrier that many people erect as soon as they see an equation. Games truly can make learning fun.

Rather than present the material as it was chronologically broadcast, I have combined it into chapters with related concepts. While chapters do cross-reference each other and highlight connections between certain concepts, you are invited to skip around and drop into chapters that seem interesting, and then return where you were later. It is not necessary to read this book from front to back.

Before ending this introduction, I would like to express how 'GameTek' and *The Dice Tower* podcast have opened doors for me, and allowed me to meet many amazing people. Researching and preparing 'GameTek' is what inspired me to try my own hand at designing games, and led to my own podcast, *Ludology*, and my professorship at the NYU Game Center.

Thanks so much to you and to all the fans over the years. My life is immeasurably richer from being a part of the global community of gamers.

Chapter 1

How to Win at
Rock-Paper-Scissors

In a game where there is no hidden information and no luck, there will always exist a series of moves for one player that will lead to a win or a draw. *Tic-Tac-Toe* (*Noughts and Crosses*) is a trivial example – to guarantee a win or draw, the first player must always play in the centre square – while more complex examples like *Chess* or *Go* are presently impossible for computers to map out.

However, this is not true for most of the games we play.

In *Rock-Paper-Scissors*, unlike *Tic-Tac-Toe*, there is no single series of moves that you should always play in response to what your opponent does. If you always play Rock, your opponent will catch on and start always playing Paper.

The core concept of *Rock-Paper-Scissors* – that A beats B, B beats C, but C beats A – is a fundamental idea in games. Surprisingly, the earliest trace of the game is in writings from the Chinese Ming dynasty, in the 1600s. Compare this with dice and playing boards, which go back more than 5000 years!

From China, the game spread to Japan and the rest of Asia, and then to North America and Europe as migration carried people to those continents. The symbols and their names have changed in various cultures, but the current Rock, Paper and Scissors actually match what was originally used in China.

Rock-Paper-Scissors Strategies

One option in playing a series of *Rock-Paper-Scissors* games is to try to guess what your opponent is going to do and craft a strategy around that. For example, if your opponent always plays Rock, or plays it more than a third of the time, you switch to playing Paper more often.

This is called an 'exploitation strategy', because you're trying to take advantage of suboptimal play by your opponent. The danger with exploitation strategies is that they are open to counter-exploitation. If you switch to playing mostly Paper, your opponent may switch to playing Scissors more often. So, you switch to Rock, and around and around we go.

However, is there a way to get off this exploitation strategy merry-go-round?

Mathematicians studying game theory have proven that, for any game, there is a mix of strategies called the 'optimal strategy'. If you play the optimal strategy, it doesn't matter what your opponent does — you will end up with a consistent result.

Now, for *Rock-Paper-Scissors*, the optimal strategy is simply to play Rock, Paper and Scissors each one-third of the time. But you can't play them in sequence or in a particular order. This is known as a 'mixed strategy', where you play a suite of moves in different proportions, selected randomly.

There are several interesting properties of using this strategy in *Rock-Paper-Scissors*, and mixed strategies in general. The first is that it does not guarantee that you will win any single game. However, if you play many, many games, there is no strategy your opponent can use that will cause you to lose over the long haul.

Next, the outcome of this strategy is invariant – it does not change, regardless of what your opponent does. It factors out what your opponent does. For example, let's say your opponent plays Rock 100 per cent of the time. Now, against this strategy you will win one-third of the time, when you select Paper, lose one-third of the time, and draw the other third. In fact, against any strategy your opponent picks, you will have an equal number of wins and losses.

Of course, if your opponent is playing a suboptimal strategy, you may be tempted to change your strategy to try to exploit it – switching to more Paper if your opponent is playing more Rock, as we discussed earlier. But then you are opening yourself up to your opponent taking advantage of your play, and thereby ending up with a worse result.

So, the optimal strategy is the safest strategy to take, in that it guarantees you a certain result. It is not necessarily a strategy that helps you to win more than your opponents, or even as often as them. If all the players are playing optimal strategies, then any player who does something different gets a worse result. This is called a Nash Equilibrium, developed in 1950 by John Nash, the mathematician who was the subject of the 2001 film *A Beautiful Mind* and who first explored this area of game theory.

Finally, there is the 'Bart Simpson strategy':

Lisa: Look, there's only one reasonable way to settle this, Bart — *Rock-Paper-Scissors.*

Lisa (thinking): *Poor predictable Bart. Always takes Rock.*

Bart (thinking): *Good old Rock. Nothing beats that!*

Bart: Rock!

Lisa: Paper!

Bart: D'Oh!

How to Win at *Rock-Paper-Scissors*

A 2014 study from Zhejiang University titled 'Social Cycling and Conditional Responses in the *Rock-Paper-Scissors* Game' (described in the media as 'How to Win at *Rock-Paper-Scissors*'),[1] had 360 students play 300 games each of *Rock-Paper-Scissors*. The students earned points based on how they did: zero points if they lost a game, one point if they tied, and a set amount if they won. The students were broken up into five groups, each assigned a different amount if they won — ranging from 1.1 (only slightly better than a tie), up to 100 (way better than a tie).

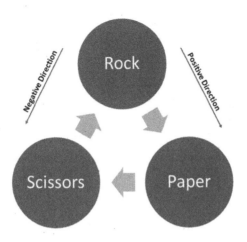

So, did the players play an optimal strategy – randomly choosing each play one-third of the time? The researchers first looked at the total distribution of Rock, Paper and Scissors. They found that each was played very close to one-third of the time, the Nash Equilibrium.

They next looked at the relationship between what participants played in one game, and in the next. And this is where things got interesting.

The researchers defined moving from Rock to Paper to Scissors as moving in the 'positive' direction (from losing to winning options), and moving from Rock to Scissors to Paper as the 'negative' direction (from winning to losing options).

To play an optimal strategy means that regardless of what happens, if players play Rock in the first round, 33 per cent should stay with Rock in the second, 33 per cent should switch to Scissors, and 33 per cent should switch to Paper.

However, the study found that if players won their match, they tended to stay with the previous choice. Moving in the *negative* direction was the second most popular option. If they won with Rock, 50 per cent stayed with Rock, 30 per cent switched to Scissors, and only 20 per cent switched to Paper.

If players lost their match, they predominantly moved in the negative direction. If they lost with Rock, 40 per cent switched to Scissors, while 33 per cent stayed with Rock, and 27 per cent switched to Paper.

And if players tied, they were slightly more likely to either pick the same choice again (staying with Rock about 38% of the time) or move in the *positive* direction (switching to Paper). Moving in the negative direction (switching to Scissors) was least likely in this instance.

The authors do not speculate on why students exhibited this behaviour. Were they trying to out-think their opponents? Or was this just an unconscious bias?

Well, in almost all cases, players were least likely to shift to the move that would lose if their opponent played the same thing again. I think that this has to do with something called 'loss aversion' – if your opponent stays the same and you lose, you will regret it more than if you changed and lost. (More on this in Chapter 6.) And this subtly pushed the players' choices.

Okay, so players in this game weren't perfect random-number generators. But can you exploit these results?

The authors constructed an optimal strategy based on what they had observed, which leads to an increase of wins from the standard 33 per cent to 36 per cent. So, the effect is real, and can be exploited, but even a sophisticated analysis only increases your score slightly. Let's say you get zero points for a loss, one for a tie, and two for a win. Exploiting this effect would increase your average score from 1 to 1.1. This may be enough if you've got a lot of money on the line, but I'm not sure that it deserves the hyperbolic title 'How to Win at *Rock-Paper-Scissors*'.

However, there is another way to lift your overall win percentage.

Rock, Paper, Scissors, Choice

In another study conducted on *Rock-Paper-Scissors*,[2] researchers at the University College London paired participants up to play a series of *Rock-Paper-Scissors* games.

In some cases, both participants were blindfolded. In others, one person was blindfolded and the other person was not. In either case, they still did the old one-two-shoot method of timing their hand-position selection.

A prize was given to the player who won overall, so there was an incentive to win.

When both players were blindfolded, the number of draws was exactly equal to what you would expect – one-third of the matches were draws, to within half a per cent. But in the blind–sighted pairings, the number of draws was above what you would expect – close to 40 per cent of the matches were draws, more than a 10 per cent increase over the expected values.

What's going on?

There is a psychological imperative to imitate what we see others do – monkey see, monkey do, if you will. This is hardwired pretty far down in our brains and can be difficult to override.

In order to imitate what the blindfolded player is doing, the sighted player must slightly delay putting their hand out. (In fact, when experimenters took out the games where the sighted player put their hand out first, or basically at the same time, the percentage of draws really shot up.)

However, this must be subconscious. First, there is very little time to react and adjust. And second, if the players really are trying to cheat by adjusting to what they see, they are doing a bad job of it. If you see that your opponent is throwing out Rock, you should quickly switch to Paper, not Rock. So, this is reflective of the brain's imitative circuits, not a higher-level function.

Interestingly too, the second-highest choices were not optimal either. For example, when the blind player played Scissors, the sighted player also played Scissors 38 per cent of the time, rather than the expected 33 per cent. But the second most commonly played symbol in this case wasn't Rock – which is what you would expect if there was conscious decision-making to beat Scissors. It was Paper, which was played 33 per cent of the time. Rock, the best choice against Scissors, was only played 29 per cent of the time.

The moral of the story is that people don't – or maybe even can't – cheat at *Rock-Paper-Scissors*. The blindfolded players actually won more than the sighted players. So, next time you play, you might try closing your eyes as an advanced strategy!

Biofeedback: Count on Your Heart to Trust Your Gut

There is a famous experiment called the Iowa Gambling Task.[3]

In the Iowa Gambling Task, players get a starting stake of money and are presented with four decks of cards. Each turn, they pick a card from any of the decks and it tells them how much they gain or lose. The trick is that the decks are biased, but the players don't know that. Some are very high-risk, some are low-risk, some are high-reward, and so on. Most players take about 50 cards before they start to draw only from the decks that give them the best results. And it often takes another 30 draws before they realise they are doing it and can logically explain their actions.

Recently, experimenters used the Iowa Gambling Task to gain a deeper appreciation of how we learn about the world. The experiment was twofold. First, the experimenters measured the body's classic stress reactions during the game – things like pulse rate and sweatiness of palms. It turned out that after only ten card draws, the stress reactions increased when drawing from a negative deck. The unconscious had already started to associate certain decks with bad results, even though this information had not entered the conscious brain yet.

The experimenter also asked the players to count their heartbeats while they played. Those who were better at that task also switched to the good decks sooner, and did better at the card game overall. The researchers hypothesised that those players who were more in touch with feedback from their bodies noticed the stress signals sooner and were able to bring subconscious information up to the conscious brain faster.

So, while you shouldn't trust your body's imitative circuits to win you a blind—sighted *Rock-Paper-Scissors* tournament, it could mean the difference between winning and losing in other games. As gamers, we should learn when to listen to our bodies and gut reactions. If you have a bad feeling about something, there may be a reason for it. But our brains are also supremely capable of seeing patterns where there aren't any. I'll have a lot more to say about that in Chapters 7 and 8.

Chapter 2

What Makes a Game?

Buddha's 'Forbidden' Games

One of the things that fascinates me about games is that they are such a fundamental part of human endeavours. *Mancala*, *Backgammon* and *Go* all date back thousands of years. The earliest known set of dice was part of a 5000-year-old *Backgammon* set discovered in Iran.

So, I am always on the lookout for historical information about what games were being played by different cultures throughout history. One such source is the *Brahmajāla Sutta*, which is a Buddhist text, recording the teachings of Buddha, most likely written down in the 5th century BC. For some context for those of you more familiar with European history, this is around the time of the war between Greece and Persia, which featured the battles of Marathon and Thermopylae, the latter of which was featured in the movie *300*.

The *Brahmajāla Sutta* describes what should and should not be done by the ascetic or the recluse. It includes a long list of activities that should not be undertaken by a Brahman attempting to achieve enlightenment, including attending certain types of theatre, growing and eating foods in certain ways, even purchasing types of couches and furniture. And smack in the middle is a discussion of games.

Now, some have called this the list of 'Buddha's Forbidden Games'. I'm not a Buddhist scholar, but, having read the surrounding context, I don't see it this way. It appears to specifically apply to monks and recluses. The section opens with this: 'Whereas some recluses and Brahmans, while living on food provided by the faithful, continue addicted to games and recreations.' It then lists eighteen specific games and activities, among which are race-style games like *Ashtapada* and *Dasapada*, dice games, ball games, stacking games, games like *Hopscotch* and *Charades*, and various other activities, such as doing somersaults or playing with toy windmills.

The section ends with this: 'Gotama the recluse holds aloof from such games and recreations.'

This is fascinating, not just to see what games and activities existed during this period, but also the attitude that is taken toward them. This is a very specific list. Does that mean that there are games not on the list that are considered acceptable and elevating to the spirit? Certainly, through history, games such as *Go* and *Chess* have been held up as examples of cultured activity.

Athenian Liturgy

Games have been used to predict behaviour, find better postal routes and even, as it turns out, to build ancient taxation systems. One such system is the 'liturgy', or *leitorgia* in Greek, which was first developed in ancient Athens as a way of ensuring the upkeep of the city state.

The liturgy system relies on the wealthiest people to give money to the government for specific tasks. Think of it as a sponsorship. For example, if you had a 'trierarchy' liturgy, you were responsible for equipment and maintenance of a trireme (warship) for a year, and were also expected to work as the ship's commander.

The government assigned liturgies to different citizens based on the needs of the city state and citizens' presumed wealth. Although some citizens really took their liturgy to heart and fulfilled it with pride and a sense of civic duty, as in our own times – and probably under any taxation system ever – most did it begrudgingly.

Now, there was no formal way in Athens for determining who owned what, or being able to tally someone's wealth. Records just weren't that sophisticated, and the effort and compulsion required to enforce such a system was not feasible for the resources available.

But people's fortunes rise and fall over the years. How did the government identify who exactly the richest people were each time a liturgy was required?

Before we get to these answers and their relation to game theory, I would like to point out a few interesting side effects of this system. First, people tended not to be very ostentatious with their wealth. The more people thought you had, even if you didn't, the more expensive liturgies you would be ordered to fulfil. Also, there were many techniques for hiding wealth: transferring it to other people, keeping assets in other cities, and so on. The parallels with offshore accounts and other tax shelters we see today are striking.

So how did the Athenian government make sure they correctly identified the wealthiest people? They instituted an ingenious 'challenge' process. If you were assigned a liturgy, but believed there was someone else with more wealth than you who was not also assigned one, you could challenge them.

Here's how the process went.

First, the Challenger named the person they believed had more wealth than they did. Let's call them the Defender.

The Defender had two options. First, they could agree with the Challenger and say, 'Yes, I do have more money than the Challenger, and I will take on the liturgy instead.'

Or, if they believed the Challenger actually had more money, they could refuse to take on the liturgy. In this case – and here's the fun part – the two parties exchanged all their possessions. Anything that belonged to the Challenger was given to the Defender and vice versa. House, land, furnishings, money held by bankers – everything. Then the Defender had to pay for the liturgy with their new property. This exchange was called *antidosis*, which is Greek for 'giving in exchange'.

I find this unbelievably cool. It is like a giant game of *Chicken*. Sadly, there are no recorded instances of an *antidosis* actually taking place, but I absolutely love the concept. It is very reminiscent of the 'I cut/you choose' mechanic in games like *New York Slice* or when making siblings share cake, but for much higher stakes.

'Gaming' the System

During the 2012 London Olympic Games, there was a scandal about, of all things, badminton.

Four teams – two from South Korea, one from China and one from Indonesia – were ejected from the Olympics due to intentionally throwing a match. In at least one of the matches, both teams were obviously trying to lose, which was pretty funny to watch. The crowd was booing, the officials gave out several warnings, but it just continued. Then after the matches were mercifully completed, Olympic officials decided to expel the offending teams.

So why would you go all the way to the Olympics and then throw a match? Did they stand to gain lots of money from badminton bets if they lost? Did the other teams have compromising photos of them? Sadly, the answer is a bit more prosaic. They did it to help them win – not the game, but the gold.

Badminton is organised in a fashion similar to many other sports, like soccer. There is first a 'round-robin' play, where teams play a set schedule, win or lose. Then the top teams from that round advance to an elimination round, and their opponents are allocated based on how they performed during the round-robin.

Due to some upsets, in 2012 some of the highest-ranked teams toward the end of the round-robin were ones that were usually considered 'weaker', based on their past performance, while some of the powerhouse teams were a little lower in the standings. This meant that some teams got into the situation where, no matter whether they won or lost, they were going to advance to the next round.

But if they lost, they would get to play an easier opponent in the elimination round. Losing their final round-robin game gave them a better chance of winning the tournament overall.

So, the question becomes – what is the 'game' here that they were trying to win? Was it the individual matches? Or was it the overall tournament? In my opinion, these teams were within their rights to lose those matches. Their goal was to win the gold. The individual matches are a means to an end.

There are many board and card games where there are a series of 'matches', and ones that you may want to deliberately lose. In *Bridge*, for example, you could say that each trick is a game. And there are many times that you will deliberately lose a trick to set up a winning play later in the hand.

The fault here is with the design of the tournament. If the tournament is set up in such a way that playing to lose is advantageous, you can't fault the teams for doing just that. As a game designer, you can't dictate the approach your players will try to take.

Badminton, like many sports, does have a rule that says, 'not using one's best efforts to win' is illegal. And many other sports have similar rules. But it is well known that, at the end of the season, football teams that have a secure place in the playoffs will bench their best players for rest and to avoid injury. And there has long been speculation that in basketball the bottom teams throw games to get a better draft pick for the next year.

Similar situations have arisen in soccer tournaments, which also have the round-robin-followed-by-elimination format. There have been a few examples of teams not playing to win because of the current standings.

As my son likes to say, 'You play the game you're given, not the game you want to have.' Strategies that are against the 'spirit' of the rules, but not the letter, are fine by me. So, in this case, I don't blame the players. They are merely playing the game they are given. It is the tournament organisers who should shoulder the blame.

Speaking of playing the game you are given, 2016 saw the election of Donald Trump as the 45th President of the United States. It was a victory that many claimed was a failure of the US voting system, as the candidate who won the 'electoral college' did not win the 'popular' vote (that is, less than 50 per cent of votes were for the winning candidate). Others put blame on supporters of Bernie Sanders or Jill Stein for 'splitting' the Democratic vote.

This is not unheard of. US presidential elections have been prone to the 'spoiler' vote on a number of previous occasions. Ross Perot in 1992 and Ralph Nader in 2000 both had a major impact in siphoning votes from other candidates (George Bush Jr and Al Gore, respectively) and shifting the results of these elections.

A better system is a ranked voting system, such as that used in Australia. In this case, voters can rank their options in order of preference. In most ranked voting methods, it is perfectly acceptable to leave out undesired choices – it is assumed these choices are tied for last. This gives voters the most flexibility.

There are a bunch of methods for resolving ranked votes. The simplest is called 'instant runoff voting'. With this method, you toss out the candidate who received the fewest first-choice votes and award those votes to the next-highest ranking candidate on the ballot. Repeat until one candidate receives more than 50 per cent of the votes.

In addition to preventing the spoiler effect, this method, if implemented in US presidential elections, would also dramatically enhance the viability of 'third-party' candidates. I am sure there are many people who would vote for a third-party candidate, but feel that they would be throwing their vote away. If they knew that, in the case that the third-party candidate did not receive enough votes, their vote would go to their next-choice candidate, many would be more likely to express their preference for so-called 'fringe' candidates. It may even give third-party candidates a true chance at taking the White House – which is why the Democrats and Republicans have no interest in changing the system. However, it has started to get some traction in the United States. In 2018, the state of Maine adopted ranked voting.

I'll finish this discussion with one last voting system from recent history: a rather convoluted system used to elect the

Doge (or Duke) of Venice from 1268 until the end of the Venetian Republic in 1797.

First, 30 electors were chosen at random and were reduced to 9 by drawing lots. These 9 people then elected 40 people. Those 40 were reduced to 12 by drawing lots once again. This group of 12 then elected 25 people, and then lots were drawn once again to get down to 9. These 9 elected 45 people, who were reduced to 11 by lots, who elected 41 members, who got to elect the Doge.

The idea was to make corruption in the system very difficult.

Venice, elections, convoluted rules … it's a game waiting to happen.

Dice, Gambling and Star Systems

Dice are a ubiquitous part of games. They are perhaps the single iconic image that represents games in our consciousness and they have an incredibly long history. The earliest dice found are almost 5000 years old, and the Egyptian games of *Mehen* and *Senet*, which date back almost 6000 years, used throwing sticks as a randomising element.

Throughout history, dice games have had a negative reputation. Dice have frequently been considered a vice, and not just in the list of Buddha's forbidden games.

In ancient Rome, gambling was illegal but endemic. If caught playing a dice game, the players had to pay a fine of four times the stakes. To make ascertaining the stakes more difficult, the citizens started using tokens to represent the money they were playing for – they basically invented casino chips. Cheating was also rampant, as people made dice with lead inserts, among other devious devices. A variety of anti-cheating devices were developed, including one we still use today – the dice tower, which prevents skilled dice throwers from controlling how the dice they roll will land.

Ancient Greeks had a complex relationship with dice games. Many philosophers came out against gambling in general, and dice games in particular, worried that they would cause chaos in society. However, like in ancient Rome, gambling was very popular, and there's no evidence that there was any tight governmental control.

The playwright Sophocles claimed that dice were invented during the Trojan War by the mythic hero Palamedes. The idea of heroes playing dice games during the Trojan War became a common theme in Greek art. In Homer's *Iliad*, Ajax is a fantastic warrior, but he was always the second best, behind Achilles. There are many vases preserved to this day that are illustrated with scenes of Achilles and Ajax playing dice against each other. Typically, Athena is shown in between them, her outstretched arm pointing at Achilles, to show he is favoured, and holding up fingers indicating the score for each hero. While the numbers vary, in all the vases Achilles is shown as having more points than Ajax, typically four to two. Ajax is shown as tense and coiled, and Achilles is relaxed and leaning back.

When a hero was killed at Troy, the custom was that his armour would be passed to another deserving hero. After

Achilles is killed, Ajax assumes that, since he is now the best warrior for the Achaeans, he will receive the Armour of Achilles, which was blessed by the gods. However, Odysseus takes the floor and, with the help of his silver tongue and the favour of Athena, convinces everyone that he deserves the armour instead. Ajax is furious and ends up going mad. He spends his entire life as second best.

Poor Ajax just can't get a break.

One last stop in our tour of the ancient world. The *Mahabharata* is arguably the most epic poem ever written. Composed in India over many hundreds of years, it reached its final form around 300AD. It has over 20,000 verses, and is about ten times longer than *The Iliad* and *The Odyssey* combined.

The story is about a dynastic conflict between two families, the Pandavas and the Kauravas, and events are set in motion by – you guessed it – a dice game.

The Pandavas have become strong, and the Kauravas realise they cannot defeat them on the battlefield. So, the leader of the Kauravas, Duryodhana, takes the advice of his devious uncle Shakuni, and challenges the Pandava leader, Yudhishthira, to a game of dice, knowing several key facts: according to custom, Yudhishthira is not allowed to refuse a challenge from another king; Shakuni is an accomplished cheater at dice; and Yudhishthira has a gambling problem. Sure enough, all of these factors come together in a dramatic sequence as Yudhishthira gambles away his entire kingdom, then his brothers, then his own freedom, and finally his wife.

The rest of the story is about how Yudhishthira spends 12 years in exile, restores his kingdom, and ends with a suitably epic war, but the *Mahabharata* is clear in its moral stance: dice games are not a good way to spend your time – especially if dice are your vice.

The classic dice is, of course, the 6-sided cube. *Dungeons & Dragons* popularised four other dice: the 4-sided tetrahedron, the 8-sided octahedron, the 12-sided dodecahedron, and the 20-sided icosahedron. (*D&D* also popularised the 10-sided dice, but this is not a regular polyhedron – the angle between the faces is not always the same.)

These five special shapes have been known for thousands of years, and have always been treated with mystical reverence. The first group that we are aware of who studied these shapes in detail were the ancient Greeks. Pythagoras did much of the early work, and the great Euclid (the inventor of Euclidean geometry) was able to prove that these five are the only five solids that can be made where each side is the same regular polygon. The cube, of course, has squares on the sides, and the dodecahedron has pentagons. The other three each have equilateral triangles for sides.

However, they are perhaps most identified with the philosopher Plato, who associated each of them with one of the basic elements. The cube represented earth, due to its solidity. The tetrahedron was fire, the octahedron air, and the icosahedron water. Finally, the dodecahedron was associated with the heavens. Because of the inclusion of these shapes in his philosophical writings, they are collectively referred to as 'Platonic solids'.

The Romans were also big fans of the Platonic solids, especially the dodecahedron. There are dozens of examples of hollow bronze dodecahedrons with various decorative elements and openings on different sides. They really are quite beautiful, but, unfortunately, no one knows their purpose. One proposal is that they were dice, but we can't be sure.

Let's fast-forward a long way to the 1500s and mathematician and astronomer Johannes Kepler. Kepler subscribed to the belief of Plato, Pythagoras and Aristotle that the cosmos could be explained by math and geometry, by ratios and intervals, all combining in the 'music of the spheres'. He was a firm believer in the new Copernican model of the universe with the sun at the centre, and had the advantage over the ancient Greeks of much better data for the orbits of the planets.

And so, with the hubris of a man in his early 20s, he set out to unlock God's divine plan for the cosmos. On 9 July 1595, he was teaching a geometry class and drew a circle with a triangle inside, and another circle inside the triangle. He noticed that the ratio between the two circles was about the same as the ratio between the orbits of Jupiter and Saturn. He tried many other two-dimensional geometric figures, but none of them worked properly. Finally, he looked at the Platonic solids. And by nesting the solids one inside the other, interspersed with spheres, he was able to match the known orbits of the six planets.

He triumphantly published his results in his book *Mysterium Cosmographicum* and thought he had unlocked the key to the divine construction of the cosmos.

As he wrote of his construction:

> It is amazing! Although I had as yet no clear idea of the order in which the perfect solids had to be arranged, I nevertheless succeeded in arranging them so happily that later on when I checked the matter over I had nothing to alter. Within a few days, everything fell into place. I saw one symmetrical solid after the other fit in so precisely between the appropriate orbits that if a peasant were to ask you on what kind of a hook the heavens are fastened so they do not fall down, it will be easy for you to tell them.

And yet Kepler did not rest on his laurels. He continued to work with the new, more accurate orbital data as it was collected. And he discovered a problem with his beautiful, perfect model of the universe. The planets did not move in circles, following the spheres that nested between his carefully placed Platonic solids. They moved in ellipses. Ugly, ungainly, askew ellipses.

And this is where Kepler elevates himself. He did not ignore the new data, nor did he cling to his model. It was still beautiful, but he knew it to be wrong.

Instead, he worked to relate geometry to the motion and ratios, and he eventually theorised that the planets sweep out equal areas in equal time as they swing around the Sun on their oval journeys. And, unlike his Platonic solids model, this one was right, and would eventually be used by Isaac Newton to unify all the motion in the cosmos in his theory of gravity.

Kepler, when confronted with the crumbling of what he considered the crowning edifice of his life, did not go into denial, did not cling to the past. He put the evidence first, above any model, no matter how beautiful, and was a key force propelling us in our quest for knowledge.

Chapter 3

Game Theory

The Prisoner's Dilemma

Game theory is a branch of mathematics. Its problems present players with a series of two or more choices, and a gain or loss for each choice. You then try to calculate what the best strategies are for each player.

Game theory has been applied to economics, psychology, evolutionary theory, politics, nuclear deterrence – I am sure a lot of you remember the NORAD (North American Aerospace Defense Command) computer WOPR (War Operation Plan Response) from the movie *WarGames* asking, 'Shall we play a game?'

Around 1950, perhaps the most famous problem in game theory was formally developed (although a version of it was proposed by the philosopher David Hume back in the 1700s[1]): the Prisoner's Dilemma.

This is the classic description:

Two suspects are arrested. The police visit each of them separately to ask if the other was involved in the crime, for which the sentence is up to ten years in prison. If one rats out the other, and the other remains silent, the betrayer goes free and the silent accomplice receives the full ten-year sentence. If both stay silent, both prisoners are sentenced to only six months in jail on a minor charge. If both betray the other, each receives a five-year sentence.

So, each prisoner must make the choice whether to betray the other or remain silent, without knowing beforehand what action the other prisoner will take.

The dilemma is this: how should the prisoners act?

The two choices – staying silent or betraying the other prisoner – are usually referred to as 'cooperating' or 'defecting' respectively. Now, the interesting part about this scenario is that, in one sense, defecting (betraying) is a dominant strategy. If you both defect, you are even with five years. But if you defect and she cooperates, you get away scot-free. So, either way you do equal or better than your accomplice.

But if both players defect, they end up with a worse result than if they both cooperate.

The Prisoner's Dilemma has been used to analyse everything from ecology to Armageddon. But it also crops up in board games.

Diplomacy and the Prisoner's Dilemma

One game that features a Prisoner's Dilemma right on the first turn is *Diplomacy*. In that game, each player represents a European power, and writes down their orders for their armed forces after negotiating.

	Stay Silent (Cooperate)	Betray (Defect)
Stay Silent (Cooperate)	½ ½	0 10
Betray (Defect)	10 0	5 5

Sentence (in years) players receive depending on their choices

There are some spaces on the board called 'supply centres'. The objective of the game is to control these. Two of the players (Italy and Austria-Hungary) have two supply centres that border each other. So, on the first turn, each of those countries needs to decide if it will move its unit away from the common border toward neutral supply centres, or into its neighbour's home supply centre.

If both cooperate, they do well, and if both defect, they do poorly. However, if one cooperates and the other defects and takes the other's home supply centre, it can be devastating to one and a boon to the other, allowing expansion beyond the usual borders.

Diplomacy – like most board games that use the Prisoner's Dilemma – is an 'iterated' as opposed to a 'one-shot' Prisoner's Dilemma. This means it is repeated multiple times between the same players.

It is typically assumed that defecting is a superior strategy in a one-shot game, since you will always score at least as many points as your opponent. However, the iterated game introduces a new wrinkle. Can you review your opponent's past behaviour to predict what they will do in the next round?

One reason why Austria-Hungary and Italy almost always cooperate on the first turn is that, even though the defector may gain a major advantage, he also makes a vengeful enemy very quickly. But cooperating early can set the stage for a game-long alliance, as each turn's cooperation depends on the bonds of trust.

In the mid-1980s, political scientist Robert Axelrod organised a tournament where people submitted computer programs to play an iterated Prisoner's Dilemma. The programs were duplicated multiple times and pitted randomly against each other, so it was possible for a program to play itself.

It turned out that the winning program was also the shortest. It was written by Anatol Rapoport, and called 'Tit-for-Tat'. Unlike most programs, which were complex, kept long histories of trends, and tried to outguess their opponent, Tit-for-Tat had only four lines of code, and the algorithm was dirt simple:

Round 1: Cooperate

Every other round: Copy what the opponent did in the last round

So, if the opponent defected, Tit-for-Tat defected on the next round. And when it faced itself, it would just cooperate every round, maximising its score.

In a very simple way, Tit-for-Tat almost encapsulates a moral response. If the opponent cooperates, continue cooperating. If they defect, punish them in the next round by defecting as well.

In fact, all the winning programs had several characteristics in common:

- They were nice: they were never the first to defect.
- They were retaliatory: once defected upon, they exacted revenge.
- They were forgiving: they would always attempt to return to cooperation.

This simple game reveals a profound mechanism for how what we all consider positive character traits could evolve and develop in social organisms. A few cooperators in the social group using the very simple Tit-for-Tat strategy will prosper and eventually become dominant.

Acting unselfishly can arise from purely self-interested behaviour. The way these simple rules can result in such beautiful complex strategies and behaviours is one of the things that I personally find so attractive about board games.

Retaliation and Reputation

Almost all negotiation games feature this element of trust and betrayal. And we can learn from Tit-for-Tat, where judging your ability to retaliate in the future can help guide your decisions.

For example, in *Mall of Horror*, players secretly vote to decide who will get thrown to the zombies. There is a lot of negotiation, deal-making and backstabbing. But backstabbing early in the game can get you into trouble, as there are plenty of opportunities for opponents to get back at you.

A key to these games is not to randomly renege on deals, but to save the reneging until the time when it really matters, and your opponents will not be able to respond in any meaningful way that prevents your victory. To paraphrase Machiavelli, when you defeat someone, defeat them so thoroughly that they cannot come back after you.

That's also why having a random number of rounds in the iterated Prisoner's Dilemma is important. If you know when the final round is, you should always defect on that round, since there is no opportunity for retaliation.

However, there is one other thing we need to consider: 'metagaming'. Will your opponents hold your betrayal from your last game against you in the next? Almost all gamers think this is not appropriate – it's just a game, after all. But I think we've all been in situations where this has happened, especially if you play more than one game in a row.

And while the 'You screwed me over last game, so I'm not going to trade with you this game' mentality is considered bad form, realising that, in most games, someone tends to go back on their deals toward the end, and using that knowledge to your advantage, is considered good and shrewd play.

You can see, then, how moral overtones begin to creep into decisions of if and when you should defect. And that is one reason *Diplomacy* is rightly classified as a vicious game which, if not played in the right spirit, can ruin friendships. People tend to get reputations in their gaming group. So, pay attention to yours, and play against type to keep them guessing.

The importance of having a good reputation in our social group is deeply hardwired into us. We want to be accepted by the group. We want people to like us. If we go against the interests of the group, our reputation suffers. If Austria-Hungary defects and betrays Italy in *Diplomacy*, other players are going to be less trusting of Austria-Hungry. If the behaviour continues, that player may ultimately be ostracised or ganged up on during the game.

Of course, groups can establish rules for radically different treatment of group members and outsiders. For example, an alliance of players in *Diplomacy* might betray other players – enemy players – but would not tolerate betrayal within the alliance itself. This behaviour also probably goes hand in hand with our tribal ancestry. We can be very much 'us versus them'.

People discovered very quickly that, when folks are allowed to be anonymous on the internet, they do things they would never do if they had to suffer the reputational consequences. To combat that, reputation has been harnessed as a trust mechanism by sites like eBay and BoardGameGeek (an online forum, database and review aggregator for board games), where stars, badges and thumbs-ups all help members gauge whom to trust and associate with.

These systems have worked to varying degrees. In 2010, Blizzard announced that it wanted to add a 'real name' feature to their forums, where you would have to use a verifiable name to participate, as a way to reduce trolling and anti-social behaviour. Ultimately, though, it was highly unpopular with players and Blizzard removed the requirement, because of concerns players might be harassed outside of the game. In 2012, Riot added a reputation system to their online game *League of Legends* to make players act in a more prosocial way. This was widely considered a success, as it associated a rating with players' online personas only. It is possible for bad actors to start new accounts to avoid a bad reputation ('smurfing'), but players who do that lose access to the characters, equipment and bonuses they have earned from their original account.

But reputation systems – particularly those in the real world rather than the internet – are inherently limited by the size of the group. The maximum number of people that we can actively engage with caps out at about 150 people – this is called 'Dunbar's number'. Once a group gets above this number, the capacity for members to keep track of everyone's reputations begins to break down, and subgroups form.

But here's something for you to ponder. Do games allow us a way to escape, if only for a while, from the reputation system? I can play *Diplomacy* and lie and stab people in the back, but I play with the assumption that those actions are bounded and contained within the confines of the 'magic circle' of the game.[2] Is this true? If I pull off some dastardly double-cross, will people start to wonder what I'm capable of in the real world? Or can we really create this little bubble and say that what happens here won't affect our reputations later?

Chapter 4

Memory, Choice and Perspective

If a game has hidden information that is revealed at some point, but then hidden again, is it okay for a player to keep written notes to keep track of it?

The go-to game for this question is, for some reason, always designer Reiner Knizia's *Tigris and Euphrates*. In *Tigris*, players earn different colour cubes, and at the end of the game you see which colour you have the fewest of – and that's your final score.

When you earn these cubes, they are placed behind a screen, so you don't know how many each player has.

However, if you have a good memory, it is certainly possible to keep track of exactly how many cubes of each colour each player has, especially when there are fewer players. So, some say that since it is theoretically trackable, it should be okay for a player to sit there with a pad and write down what everyone has. Or simply play without the screens.

I want to tackle this from two directions. First, should this be okay? Second, why do people want to do this?

*

Let's take things to an extreme for the first question. Let's look at the game *Memory*, where you flip over pairs of cards looking for matches. No one would say that you should take notes in *Memory*. The whole point of the game is to exercise your memory. It's right there in the title. If you write down what the cards are, it's pointless and against the intent of the game and the intent of the designer, whoever that was.

That's the key phrase: the intent of the game and of the designer. What was the designer's point? Absent specific feedback from the designer, that is impossible to know. Therefore, we need to play with the rules that come with the game. These are the ultimate representation of the designer's intent. And the default position is that players should play with the materials in the game. Anything else is changing the rules.

In *Tigris*, you place the cubes behind a screen. It's obvious that the designer intended the count to be secret. *Tigris* doesn't come with pads. A game that *does* come with pads is *Clue/Cluedo* (as it's known in different places). Players are encouraged to take notes on what is said, rather than try to remember.

That's the final answer, then. If a game comes with a pad or a whiteboard, then you can keep track of hidden information. Otherwise you can't. End of story.

Of course, it's your game and you can play it however you like – but don't involve me in your twisted schemes.

*

The second question is more interesting. Why do players want to track hidden information in some games but not others?

There are several characteristics of games where players believe they should be allowed to track information. First, memory can't be the key aspect of the game. In *Tigris,* there is a gameplay reason to keep things hidden: to prevent excessive 'analysis paralysis', where players are literally paralysed in deciding because of the number of options available to them. But keeping things hidden doesn't feel like the core of the *Tigris* gameplay experience.

Another characteristic is that the number of things that you need to remember is on the edge of what is possible. In this case, it seems to bother folks that another player, who has been blessed with a better memory, is able to be a better player because of this skill. If it's not possible for anyone to remember everything, it doesn't come up as an issue.

But why is memory any different from any other skill? Why do people think this skill deserves to be handicapped for those with poor memories? If I can quickly calculate probabilities in my head, it may make me a better player in certain games. But should other players be allowed to bring calculators? No, of course not. If some players have better memories, that's a skill they bring to the table, just like any other skill.

Improving Your Rapid Recall

Memory is a fascinating topic, and one that continues to be actively researched. And there are many games that use memory as a game element. Some of the earliest games we play as children involve memory — even 'peek-a-boo', perhaps the first game played with babies, is essentially a memory game.

There are several types of memory storage in our brains. The most prominent division is between short-term and long-term memory.

Short-term memory is where we briefly remember things that just happened. If something is important enough, like a traumatic event, or if we repeatedly learn it or see it or do it, eventually it will be committed to long-term memory.

There have been a lot of experiments about short-term memory, and most show that you can typically remember about five to nine things for about 20 to 30 seconds. The number you usually hear is that people can remember seven things, but recent studies have pushed that number down to four or five.

One of my favourite short-term memory games is *Rapid Recall*. It's out of print now, but it's pretty easy to make your own version with a pen, a pad, and a bucket of poker chips or coins.

Here is how to play.

Two people are on a team. When a team plays, one person gives clues, and the other person is the guesser. The clue-giver is given a sheet of paper from the other team that lists ten words — very simple ones, like 'ball' or 'kitchen'.

Then the clue-giver has 60 seconds to give clues to the words. They can say as much as they like without saying the actual word. For example, for 'kitchen', you could say, 'The room in your house where you cook food.' When the guesser thinks they know the answer, they don't say the word – they just throw a chip into the pot.

After all ten words, or 60 seconds, the clue-giver stops giving clues and the guesser has 60 seconds to name as many of the ten items as possible. They get one point for each word they can remember.

It's fascinating to play this game, and to watch folks play it. It's really hard to remember all ten items. I've remembered all ten maybe once or twice in all the times I've played it. And I've seen many, many people just completely blank and forget every single word. I'd say that most people get four or five, so that ties in nicely with experimental results.

An interesting aspect of *Rapid Recall* is that you don't say the words out loud. Once, as a family experiment, we played it so that the guesser *did* say the word out loud during the clue-giving step. And it typically led to people getting about two to three more answers right.

There are different ways that information gets into our brains – reading, hearing, doing – and the more ways it gets in, the better our retention is. So, by restricting players to just a single source of input – figuring out the word in their head – *Rapid Recall* makes a tough task even tougher.

There are many strategies that you can use to improve your short-term memory. Simply focusing and removing distractions can help. And people will get better at *Rapid Recall* over several rounds just with practice.

But there is another technique called 'chunking'. If you can combine multiple pieces of information into a single entity that you already know, three or more separate items suddenly become a single item.

For example, here's a string of 13 letters:

F I O Q L T U K A X R G J

Cover them up and try repeating it back. How many can you remember?

Now here's another string of 13 letters:

C A T D O G H A M S T E R

Can you cover and repeat that sequence? I'm sure you can. Because you only have to remember three things – CAT, DOG, HAMSTER – instead of 13 unrelated things. Plus, the three things are in the same category (animals), which helps you remember. If all the words in a game of *Rapid Recall* were in the same category, I think people would do much better.

If we go back to what we discussed at the beginning of this chapter, this brings up another interesting question. What actually counts as 'cheating' in a memory game?

In *Space Cadets*, players basically play a 16-tile memory game during the Tractor Beams puzzle. Many groups struggle with this minigame as they try to remember what tiles were flipped. But other groups specifically divide up the grid so that each player is responsible for remembering only three or four tiles. Just like chunking, this dramatically lessens the cognitive load on each person and makes this part of the game much simpler for the team.

As the designer of the game, I'm not totally comfortable with that strategy, but it's totally legal according to the rules, so that's okay.

But I am always surprised at how few groups do this. Do they just not think of it? Or do they feel that it verges on cheating and so don't do it? There is definitely no hard line here about what is cheating and what is not, but the fact that this strategy is debatable illuminates the issue in a very interesting way.

The Innovation Limit

Most of us are interested in new games – in games that try to do something different. In other words, games that have the long-sought-after 'innovative mechanic'.

We applaud innovation – *Dominion* with its deck-building, *Hanabi* with its reverse hidden information.

But how much innovation can we really handle?

I have no scientific data to back this up – just experience and observation – but I maintain that for an innovative game to be successful, it needs to have exactly one innovation. One. If you have two or more major innovations in a game, it will not be a success. It will be perceived as too complex.

Let's go back to *Dominion* again. This is a 'deck-building' game – a game where you use a deck of cards, and 'build' or improve that deck by buying more cards to add to it over the course of the game.

One of the criticisms that was levelled at *Dominion* on its release was that there was nothing there other than the deck-building. But looked at through this lens, we see that this was a key decision by designer Donald X Vaccarino.

Nightfall is also a deck-building game, with a single major innovation – the ability to 'chain' related cards for better effects. What would have happened if, say, *Nightfall* had been the first deck-builder to be released? If both deck-building and card-chaining were the first exposure that someone had to those mechanics, it would be overwhelming and difficult to learn without assistance.

Cosmic Encounter was the first game to add a myriad of special powers. But the rest of the game? Simple. The combat system is a slight variation on the card game *War*.

If you look at all the popular games, you will find that they either remix existing mechanics of the game or add at most one innovation. Of course, there will be a few exceptions here and there. But I think that even experienced players can only really focus on one innovation at a time.

Terra Mystica is a board game about building villages across a multicoloured hex-grid board, except that you can 'terraform' hexes on the map to change their colour.

The way that you play on the board, with terraforming and needing certain resources to do that, is innovative in the way it is managed. But it also has 'bowls of power' where power/magic tokens get cycled as you use them. That is a mechanic that also has not been seen before. The other mechanics, like having religion tracks, are fairly common in designer games like *Terra Mystica*.

How did I cope at first with these multiple innovations in the game? For the first half of the game I ignored the bowls of power and focused exclusively on the terraforming and the systems that I understood from other games. Only after three or so turns did I start to experiment with the bowls of power. I naturally focused on only one innovation at a time. And I bet that you do too.

Managing Too Many Choices

Among the things that gamers always say they look for in a game are choices – meaningful choices specifically. That is what, after all, separates playing a game from simply flipping a coin.

When we sit down to play a game, we want to be masters of our own destiny, charting our own course through the obstacles set up by the designer and our opponents. We don't want to be railroaded along particular paths and obvious decisions.

A 2011 study done at Rutgers University showed just how much we value choice in games.[1] Researchers presented players with two boxes that had different monetary prizes. In some cases, a computer randomly selected a box for them to open. In the others, players made the choice themselves. They then saw what, if anything, they had won.

When the researchers looked at what was going on inside the brains of both groups while the boxes were being selected, but before the boxes were opened, they saw something interesting. The reward centres of the brains were activated in the group that had to make the choice about which box to open. The group that had it selected for them randomly? No activation.

The simple act of making a choice, whether it pays off or not, activates the reward centres of the brain. Games with choices are rewarding – win or lose.

So, choice is good. And if some choice is good, then more choices must be better. Right?

Pull up a seat here near the fire and let me tell you one of Aesop's Fables.

A Fox was boasting to a Cat of its clever devices for escaping its enemies.

'I have a whole bag of tricks,' it said, 'which contains a hundred ways of escaping.'

'I have only one,' said the Cat, 'but I can generally manage with that.'

Just at that moment, they heard the cry of a pack of hounds coming toward them, and the Cat immediately scampered up a tree and hid herself in the boughs.

'This is my plan,' said the Cat. 'What are you going to do?'

The Fox thought first of one way, then of another, and, while it was debating, the hounds came nearer and nearer, until at last the Fox in its confusion was caught up by the hounds and killed by the huntsmen.

The Cat, who had been looking on, said: 'Better one safe way than a hundred on which you cannot reckon.'

This story goes back thousands of years, and variants appear in many cultures. The lesson is simple: sometimes having too many choices can actually hurt us.

Remember when we talked about 'analysis paralysis' earlier in this chapter? Well, research by psychologists Sheena Iyengar and Mark Lepper has shown that if you give people more than eight to ten choices, their ability to make decisions goes down.[2]

That means a lot of things. People take longer to make a choice. The quality of their choice is lower. And they tend to regret their choice afterwards.

Baskin-Robbins famously offers 31 flavours of ice cream – one for each day of the month. Yet most people don't really consider all 31 flavours. They look for a category – something chocolatey, or with mint – and then focus their choice on those few flavours. People think they wants lots of choices but they don't.

Let's relate this to games. I think that too much choice actually harms a game. For example, in *Dominion*, you win by building a deck out of cards purchased during the game. On any turn, there are about 16 cards to choose from, though realistically there are usually only about eight you can afford. In the box, there are about 30 different types of cards, and the game randomises which ones you play with each time. If you could purchase from all 30 types, it would be a worse game for that.

In the game *Struggle of Empires*, designed by Martin Wallace, there are 30 or 40 special ability tiles that you can acquire. These are laid out at the beginning of the game, and are available right from the start. For the new player, this can be overwhelming and is a weakness in the game in terms of attracting first-timers. There are just too many choices.

Now, in both games, you can train your brain to artificially reduce the number of options available to you. In *Dominion*, you might focus on only a couple of cards, and create a strategy out of them. In *Struggle of Empires*, once you play a few times, you start to see which tiles are good early, and which are good later, and which work with certain strategies.

In both games, once you have a strategy, your brain puts the various options into 'chunks', so, just like with chunking memory, you no longer have to evaluate too many different options. You bring it back down to a reasonable number.

So, having a strategic framework when you play has the added benefit of limiting the number of choices you have to consider. Therefore, you are more likely to make the best move within that strategy, as opposed to looking equally at every possible move.

Another Point of View

Another technique for making better decisions is to physically change your perspective.

Back in 1999, a team from the University of California published a research paper called 'Interactive Skill in *Scrabble*'.[3]

In *Scrabble*, players each have a rack of seven tiles, each tile with a letter on it. Players take turns placing tiles from their rack onto a board to spell out words in a crossword arrangement using letters on the board. Rarer letters are worth more points, and playing all seven tiles – or using special spaces on the board – can net even more points.

The research team wanted to investigate how the tile rack affected player's abilities. They created two sets of seven *Scrabble* tiles within which players had to create words (they weren't required to place them on the board). One rack had easy-to-find words, and the other was more challenging.

Each participant got one of the two racks of tiles. Then they were either told that they could move the tiles around, or that they could look but not touch. They were given five minutes to come up with as many words as possible from the letters on their rack.

The researchers wanted to see if physically moving tiles around helped the thought process of the students. Would it help them to come up with more words?

The results were mixed. On the simpler rack, the people who moved tiles around actually found fewer words than those who just looked at the rack – 18 words with hands versus 22 with no hands. But for students who had the 'difficult' rack, moving the tiles around definitely helped. On average, 16 words were found without moving tiles, but 23 words were found when people could move the tiles – that's almost 50 per cent more.

The researchers theorised that the result found on the difficult rack was due to something they called the 'rubber-band effect'. The idea is that, even though you move the tiles around in your head, every time you look back down at the rack your brain 'snaps back' to the tile order you see in front of you. It requires physically rearranging the tiles to help your brain see other options. With the simpler rack, it is easier for your brain to snap onto possible words, so the rubber-band effect does not come into play.

I think this applies to many game situations when we're not sure what to do. Back when I had more free time and was playing wargames, I would make it a point to frequently – every turn in some cases – get up and walk around to other sides of the map and look at the situation from another angle. Yes, the pieces were all in the same places, but the change in perspective very often showed me different opportunities or threats, and would give me new ideas to try.

Even simpler games can benefit from this. I find that, especially in a game that is new to me, I will often focus just on what's happening in the part of the board that is closest to me. I've got enough going on just figuring out my own stuff without worrying about what everyone else is doing. Moving around can help quite a bit to get over this narrow focus.

Same thing with changing the order of your hand of cards; moving pieces around on your sheet; or even playing a game with a different coloured player piece.

It isn't just mentally 'chunking' things together that can improve your gameplay. Physically changing your perspective can also enhance your game.

Chapter 5

Playing on Your Expectations

When Will I Ever Use This?

Too many believe, as Barbie used to say, that math is hard. But just as you might 'chunk' data to remember it more easily, or physically move around tiles in *Scrabble* to avoid the rubber-band effect, there are some simple techniques that you can use in your head to improve your math and make better decisions while gaming.

One of the easiest is calculating the 'expectation value' of a choice. The definition of expectation value is simple: it is the average result that you can expect from an action.

To calculate the expectation value, you multiply the value of the outcome by the probability that that outcome will occur. Then you add up all those numbers. It sounds hard, but it's pretty easy in practice.

Let's look at a couple of examples in games.

In the terrific *Twilight Struggle*, a simulation of the Cold War, you can either play a card for its operational points, which lets you put influence on the board, or you can use the event on the card.

There is one card in the deck called 'Olympic Games'. The event has each player roll a six-sided die (remember: one die, two or more dice). The person playing the card gets +2 to their roll. And the high roller gets two victory points, with ties being re-rolled. Or you can play the card for two operational points.

So, what's the expectation value for this event?

Let's take it one step at a time. The first question is, what is the chance that you will win the Olympic Games? What is the chance that you will roll higher than your opponent when you get a +2?

Later in this chapter, I will cover probability and estimating, and how you can better calculate these, but your intuition should be good enough here. Pause for a second and come up with a percentage.

What did you come up with?

The correct answer is 81 per cent. But if you were anywhere in the 70s or 80s, your answer is close enough. Most people guess around 75 per cent.

With no modifier to your roll of the dice, your odds of rolling higher than your opponent are 50 per cent, and you should intuit that the +2 should boost you a decent amount.

Okay, let's take 80 per cent as our working figure. You have an 80 per cent chance of getting +2 victory points and a 20 per cent chance of getting −2 victory points (because you will be two points further behind your opponent). So that's:

$0.8 \times +2 = +1.6$

$0.2 \times -2 = -0.4$

Add $+1.6$ and -0.4 together, and you get $+1.2$ victory points.

Even if you are math-phobic, you should still be able to do this calculation in your head. And the beauty of doing this is that now you have a really good piece of information for how you want to play this card.

If using the two operational points will net you more than 1.2 victory points, you should do that instead of playing the event. It's unlikely that you'll know for sure, but at least you'll have a good basis for making your decision.

My point is not to drain all the fun out of a game. But rather that, if you choose to use them, there are some quick mathematical techniques you can do in your head that will open new tactical and strategic dimensions for some of your favourite games. You don't need to sit down and actually calculate that there's an 81 per cent chance of winning the Olympic Games to estimate that the expectation value is a little over 1 victory point.

And, far from draining the life out of games and dragging them out, this can reduce analysis paralysis by helping you build a framework to make decisions, and it can help enrich your gaming experience by revealing new levels of understanding.

Expectation value can be a very important concept when playing a game. So important, in fact, that game designers need to be cognizant of players who use it, and take pains to make sure that the expectation value is 'fuzzy' in some way.

In *Masters of Commerce* by publisher Grouper Games (and re-released in 2012 by publisher Marabunta as *Panic on Wall Street!*), players try to negotiate deals for properties of different colours. There is a current value for each colour property, and then at the end of negotiations a die is rolled for each colour and the price may go up or down. Then players get paid for what they own. Certain colours, like red, are extremely volatile, and others, like blue, tend to vary in price very little.

The negotiation phase of the game is the fun and crazy part. Players have two minutes to scream at each other and try to negotiate prices on all the properties, and deals are up in the air until the last second.

The issue with the game, however, is with expectation value. You can look at each dice and see how many spaces the price can move up or down. Based on that, and looking at the chart, it is pretty simple – if you have practised calculating expectation value – to figure out what the expected value of each colour is.

If you do a quick mental calculation of the expected value of the colours at the start of each round, you can get a feel for what the properties are worth when the negotiations start. Invariably, the other players will overvalue some properties and undervalue others. Then all you need to do is jump onto the undervalued properties and snatch up as many as you can.

When I did this the first time I played, of course, half the time the properties ended up below expectation value, and half the time above. But with 20 die rolls over the course of the game, they ended up pretty close to average. My strategy paid off handsomely, and I ran away with the game, doubling the dollar total of my nearest competitor.

Now, I'm not telling you this to make myself look good – because it probably doesn't. It probably makes me sound like a no-fun stick-in-the-mud who drives other players crazy. But when I play a game like this, that's where my mind goes, and that's what I try to do.

If I were to play *Masters of Commerce* again, I would stick with the same strategy. In the long run, it can't be beaten. Maybe I would get more creative with the negotiations and try to put together some multi-round deals, block out some other players, and get up to other shenanigans. But in the two-minute negotiation window, there's not much opportunity for in-depth haggling, especially in a game with lots of players.

Estimating on the Fly

One of the key skills that has helped me in all kinds of situations – gaming, business, or otherwise – is being able to do rough math estimates on the fly. There are lots of opportunities to use the ability to do quick calculations to your advantage.

So, how do you improve your ability to estimate? Like all things – practise! There are a variety of common situations that will give you the chance. After you finish dinner at a restaurant, try to estimate how much the bill will be before you get it. Or next time you're buying groceries, estimate the total while you're in line.

There are also more interesting questions that can stretch your estimating muscles, and give you a chance to deal with big numbers, in more realistic situations.

For example: how fast is the Earth moving as it orbits the Sun (in kilometres per hour)?

If you want to tackle it without any hints, stop reading and give it a shot. Otherwise, here are two pieces of information that you should know:

1. The distance from the Earth to the Sun is 149.6 million kilometres.

2. The Earth's orbit is an ellipse, which means it travels faster or slower depending on the time of year. But it is close enough to a circle to use that for our estimate.

Here's another chance for you to go off and try this on your own.

Welcome back! Here's how I tackled this question.

The circumference of a circle is π (pi) multiplied by the diameter (or twice the radius): πd or 2πr. Since the Earth's radius is 149.6 million kilometres, the diameter is 299.2 million kilometres, or close enough to 300 million that we can use that. π is a tad higher than 3, so let's just triple the diameter.

Remember, we're estimating in our heads, so keep the numbers as round as possible. So the Earth travels 900 million kilometres in a year. We need to get that down to kilometres per hour.

A great tool for estimating is breaking problems down into manageable chunks (yes, 'chunking' problems also works here to overcome your memory's limits!). So, let's start with kilometres per day. We need to divide 900 by 365.

This seems like a tricky thing to estimate. But another really useful tool in your arsenal is what I call 'bracketing'. Use round numbers that are a little high or a little low, and get those answers. Then pick a number in the middle.

For example, here we need to divide 900 by 365. Well, 900 divided by 300 would be 3. And 900 divided by 400 would be ¾ or a little over 2. So, the answer is somewhere between 3 and about 2.2. Since 365 is bit closer to 400 than 300, we'll shade our estimate a bit above the halfway mark, so 2.8 is a good choice.

Now we have estimated that the Earth goes about 2.8 million kilometres in a single day. That's pretty cool! All we need to do is divide 2.8 by 24, to go from days to hours, and we're done. Dividing by 25 would be a lot simpler than 24, so let's just use that.

A last trick for you, I call the 'rule of 100s': dividing by something is the same as multiplying it by whatever would bring it up to 100 (or a multiple of 100), and then lopping off a few digits.

For example, dividing by 50 is the same as multiplying by 2 ($50 \times 2 = 100$), and then moving the decimal point across twice. Dividing by 25 is the same as multiplying by 4 ($25 \times 4 = 100$), and then dropping the last two numbers.

Dividing by 333 is the same as multiplying by 3 (333 × 3 ≈ 1000), and dropping the last *three* digits.

So earlier, when we were dividing by 365 we could have used this trick. We just could have tripled the '9' in 900 million to get 2700 million and then moved the decimal point three times, which would have given us an estimate of 2.7 million kilometres per day instead of 2.8 million. And since we're estimating, either of those would be close enough!

Back to our final step. We need to divide 2.8 by 25. Instead of dividing by 25 then, we can just multiply by 4, which, of course, is the same as doubling and doubling again. 2.8 doubled is 5.6. Doubled again, it is 11.2. Let's just call it 11 – we're estimating! Then move the decimal across twice from 11 to 0.11.

So, we go from 2.75 million kilometres per day to 0.11 million (or 110,000) kilometres per hour.

One of the great things about learning to do estimates is that you can get the precise answer at some point – whether it's by looking at the real bill, looking something up on the internet, or having your accounting department crunch some numbers for you. So, you'll be able to see how close you got, where you went wrong, and how to do better next time.

Looking up our question above, the actual speed of the Earth around the Sun is ... about 107,000 kilometres per hour. Our estimate looks pretty darn good considering we just did it in our heads!

To summarise:

- Practise!
- Learn some basic techniques:
- Round some numbers high and some low
- Use bracketing to get high and low values and go in between
- Multiplying is usually easier than dividing – use the 'rule of 100s'
- Check your estimate against the actual answer, and if you were way off, figure out why.

As you estimate things more and more, you'll get faster and more accurate. Having confidence in your ability to use math in any situation will prove to be an invaluable tool.

Flipping Probabilities

Going back to the Olympic Games example in *Twilight Struggle*, you may be wondering how I so easily calculated 'roughly 80 per cent' in my head.

I want to share a little probability tip that I use to help calculate things quickly.

Probabilities are expressed as fractions or percentages. And when all possibilities are considered, they have to add up to 1, or 100 per cent. So, if I know the chance of something happening is 25 per cent, there is a 75 per cent chance of it *not* happening, and vice versa.

So far, so good.

People often struggle with calculations that could be made much simpler by flipping them around. For example, let's say you need to roll a 5 or a 6 on a six-sided die, but you get a re-roll if you miss. How would you figure out the chances for success?

Now, one way is to figure out the chances that you succeed on the first roll, and then figure out the chances that you miss on the first roll, but succeed on the second roll. That requires two separate calculation steps and can get a little hairy. Let's see how hairy.

(Before we start, here's another quick tip for probability. If you say you need to do A *or* B, you add the probabilities. If you need to do A *and then* B, you multiply the probabilities.)

First, to figure out the chances of rolling a 5 *or* a 6, we add the chances of rolling a 5 (⅙) to the chances of rolling a 6 (also ⅙):

⅙ + ⅙ = 2/6 (or ⅓)

So the chance for success on the first roll is ⅓.

Next, you need to figure out the probability of success on the second roll. In order for this to happen, we need to fail the first roll *and then* succeed on the second. There's that key phrase 'and then'. We multiply the chance of failure on the first roll (⅔) by the chance of success on the second (⅓):

⅔ × ⅓ = 2/9

So the chance of succeeding on the second roll is 2/9.

Last, you need to figure out the chances of succeeding on the first roll *or* the second roll. Since we said 'or', we need to add the two. So, our total chance of success is ⅓ + 2/9, which is 5/9, or slightly higher than 50 per cent.

Now, that much math can be tough to do in your head. In cases like this, it is much simpler to flip the problem around. So instead of asking 'What are the chances of succeeding on the first roll or succeeding on the second roll?', ask 'What are the chances of failing twice?'

To do this, we just multiply the chance of failure on the first roll ($\frac{2}{3}$) with the chance of failure on the second roll (also $\frac{2}{3}$):

$$\frac{2}{3} \times \frac{2}{3} = \frac{4}{9}$$

The chances of failing are $\frac{4}{9}$, which means the chances of success are $\frac{5}{9}$ – which is what we figured out earlier.

This is the power of flipping the problem around.

As another example, let's look at a situation from the game *Warmachine*. In *Warmachine*, when you attack, if any two of the dice you roll match, you get a critical hit and do more damage.

If you roll two dice, you need to roll doubles, which is easy to calculate: a $\frac{1}{6}$ chance. But if you're rolling three or four dice, what are your chances of a critical hit?

Now, the direct approach here can also get hairy. You need to look at each die that's rolled, and look at the chances that it matches one of the other dice that have been rolled. There are lots of possibilities to consider.

However, it turns out that it is much simpler to figure out the chances of all the dice being different – in other words, *not* getting a critical hit.

For example, consider rolling three dice, one at a time. The first die can't match anything because, well, it's the first die you rolled. The second die has a $\frac{5}{6}$ chance of not matching the first die, and the third die has a $\frac{4}{6}$ chance of not matching either of the first two.

Your chances of rolling three different numbers are $\frac{5}{6} \times \frac{4}{6} = \frac{20}{36}$, which is about a $\frac{2}{3}$ chance, and that is as close as you need to get when you're calculating things in your head.

Therefore, you have a $\frac{1}{3}$ chance of getting a critical hit, since you have a $\frac{2}{3}$ chance of not.

If you go to four dice, the chances of not getting a critical hit are simply $\frac{5}{6} \times \frac{4}{6} \times \frac{3}{6}$ (or half of what we figured out for three dice, since $\frac{3}{6} = \frac{1}{2}$). You have about a $\frac{2}{3}$ chance of getting a critical hit.

Now, with a little practice, you can easily do these types of calculations in your head. So, when faced with a complex probability problem, try to flip it around and calculate the chances of it *not* happening.

More often than not, it will dramatically simplify the math.

Chapter 6

Feeling the Loss

My daughter went to a semi-pro baseball match on the weekend. Since it was a small team playing, they had activities between innings, and for one of these they invited a spectator out onto the field and presented her with three sealed boxes.

There was an iPod in one box and coupons in the other two. She was told to pick one box but not to open it.

Now, at least one of the two remaining boxes had to hold one of the coupons. The emcee opened one of the boxes and showed her that there was indeed a coupon inside.

The woman was then given the choice to keep the box she had chosen or switch to the unseen box.

She decided to stick with the box she had, which turned out to hold the other coupon.

My daughter had a big discussion with her friends about the choice. Did it matter if she switched or not? They decided there was definitely no benefit to switching. There had to be another unselected coupon, so the emcee didn't give out any additional information. There was no downside to switching, but no benefit in doing so. Since there were only two boxes left, the chances the fan had the iPod in her box were fifty-fifty.

What do you think? I'll come back with an answer to this at the end of the chapter.

Endowment and Endowed Progress

Before we get to the math surrounding this problem, it's important to look at the psychology behind it.

First, a quote from Ayn Rand's *The Fountainhead*:

I am the most offensively possessive man on earth. I do something to things. Let me pick up an ashtray from a dime-store counter, pay for it, and put it in my pocket — and it becomes a special kind of ashtray, unlike any on earth, because it's mine.

I think most of us understand the sentiment expressed by that character. A similar observation has been made by science, going back almost 30 years. In economic and psychological circles, this tendency to give something more value because it belongs to you is called the 'endowment effect'. It is a problem for classical economic theory, which assumes purely rational actions by people when valuing things. If something gains more value just because you have it, that throws a lot of theories out of whack. So, scientists have been trying to measure this effect in more detail, and figure out its origins.

In one experiment, for example, students were reluctant to give up a coffee mug they had been given in exchange for a bar of chocolate, even though, when presented as equal options, most students preferred the chocolate.[1] This was further explored in 2008 by a team from Stanford University, who conducted a similar experiment while doing brain scans.[2] The location of brain activity during the decision-making process suggests that the endowment effect is an emotional response tied to a sense of loss.

Interestingly, however, the effect is not seen for all types of goods. If, for example, the students are given tokens that can be exchanged for coffee mugs, instead of actual mugs, the endowment effect does not happen. (The endowment effect is also seen in monkeys, but only with food, and not with toys and other objects, which may indicate that it is a holdover from our evolutionary past, but was originally focused on things key to survival.)

In game terms, the lesson to be drawn here is not to become too attached to the assets you have as a player — whether they are territories, cards, trade goods, whatever. Keep in mind that you will tend to overvalue what you have in your possession, and, if you step back and try to do an objective evaluation, you will improve your play. You can also use this to your advantage by gaining a better understanding of the thought processes of your opponents in negotiation games.

Here's a related example. In 2006, researchers went to a car wash and gave out loyalty cards that would eventually give people a free car wash. Half the people received a card that needed to be punched eight times to get the freebie. The other half received a card that needed to be punched ten times to get the free car wash – but two spots were pre-punched. Both cards needed to get eight more punches. Theoretically, it should not have made any difference to the number of people who completed their cards and got the free car wash.

Well, only 19 per cent of the people who got the eight-punch cards got the free car wash. But 34 per cent of the people who got the ten-punch card with two punches already on it – in other words, almost twice as many people – got the free car wash.[3]

This is called the 'endowed progress effect'. The researchers theorised that once we have started on a journey, we are more likely to finish it than a journey that has not yet begun. The people who had two punches on their cards already felt invested in it. They felt a sense of progress, a sense of movement toward the goal. And this is even though they had not done anything.

Is there a game that uses this technique? Yes, there is – and it almost exactly duplicates the car wash experiment, even though it predates it by ten years. I am talking about *The Settlers of Catan*. In *Catan*, you are trying to get to ten victory points by constructing buildings and roads, but you already start the game with two buildings (and so two victory points). This gives the players a sense of accomplishment right at the start, and that feeling of building something up is maintained throughout the game.

Here's another way game designers can take advantage

of the endowed progress effect: if you have a game where you are heading toward a goal, make the first few steps along that path ridiculously simple. Give the players an immediate sense of gratification and progress.

In *Catan*, points are pretty easy to come by early in the game, but gaining them gets harder, as spots to build new towns get rarer. You need to scramble for those last few points to reach victory.

These two techniques – putting players already partway on the path to victory, and making the points needed for victory easy to get at the beginning and more difficult later on – are terrific ways for the game designer to get players involved early and keep their interest throughout the game. Most games tend to have a flat or even reverse difficulty trend to scoring victory points. Usually there is no difference in how hard it is to get points early or late in the game. In fact, I would hazard a guess that in most games it is easier to score big points late in the game than early.[4]

The Sunk Costs Fallacy and Loss Aversion

Let's say that you've bought a game. You're very excited about it, and bring it to the table, but it doesn't live up to your expectations. In fact, you *really* don't like it. But you did spend a chunk of change on the game, so a few weeks later you decide to bring it out again and once again it's a dud. You are sad. You've spent money on it, and taken the time to learn and teach it. Maybe you try with a different number of players to see if the game will finally come to life.

Is this a rational way to act? If you've spent time and money on a game, shouldn't you try everything possible to get some enjoyment out of it?

Well, no. Economic theory teaches us that, to make a rational decision, we should focus only on the future costs and benefits – not any costs that you have paid in the past. That money is already spent and can't be recovered. All that you have control over is what you will do with the resources you have now – be they time, money or anything else.

However, we now know from the endowment theory that people do not always make rational decisions. In fact, in this case, people will typically value what they've spent in the past, and will try to make good on it.

Resources that you have spent in the past are called 'sunk costs' and the tendency for people to value what happened in the past is called the 'sunk cost fallacy'.

Here's another example. You're about an hour into a new game and several things have become obvious: it's going to last at least another two hours and it's really bad and not likely to get any better. You've been through the rules and know with certainty that you've seen everything the game has to offer, and there are no surprises in store. It's a stinker.

Do you get up and walk away? Or do you spend the extra two hours to finish up, since you've already come this far? Economic theory says that you should get up and walk away. You can't get that hour back. But you can spend the next two hours doing something that you will enjoy. The fact that you've already spent an hour should have absolutely no bearing on your decision. You should simply look at where you are now, and what you can do going forward.

But many people will not walk away from a game in progress – or a movie, or a baseball game, or any other activity.

Psychologists have many explanations for why we are susceptible to the sunk cost fallacy, but most of them boil down to a thing called loss aversion. This is central to human psychology and it basically comes down to this: losses make us feel much worse than gains make us feel good.

Sticking out a bad decision in the hope it will turn around is classic loss aversion. We want to avoid the loss of the resources we've already invested – hence the sunk cost fallacy. We also tend to think that our past decisions have a better chance of leading to a good result than they actually do. We are overly optimistic that choices we've made will pay off in the future.

One of my favourite examples of this is from a 1968 experiment by Robert Knox and James Inkster.[5] They went to a racetrack and asked people who were just about to make a bet how confident they were in their pick, and then asked the same question of people who had just made a bet.

They asked people to rate their chances on a scale of one to seven. The people interviewed 30 seconds *before* making a bet averaged 3.4. Those interviewed 30 seconds *after* placing their bet? 4.8. They were much more confident about their chances, much more optimistic, for absolutely no reason other than that they had put money down on a horse.

To play the best you can, you need to evaluate the current situation as it stands and not think about what happened in the past. Look at your current assets, the board position and the other players, and figure out what the best move is as if you had just sat down at the table. Just like with avoiding the endowment effect, it's important to evaluate your position and assets based on their *current* value – not on the time you've invested or the ownership value you place on them.

It's really hard to do. And you can certainly argue that approaching each situation as if the past didn't exist can destroy the narrative thread that engages players in the game, and can make it less fun to play.

But think about the sunk cost fallacy the next time you're stuck in a less-than-stellar game, or want to bring out something that failed in the past, just in the hope that it will be better this time. Think about where you really want to spend your time – which is, after all, our most precious resource.

The Monty Hall Problem and Regret

All right, let's come back to that problem I posed to you at the start of the chapter. If you recall, a woman was given the option of three sealed boxes – one contained an iPod, the other two contained coupons. After selecting her box, the emcee revealed one of the remaining boxes contained a coupon, and the woman was given the option of changing her guess.

Should she have?

Well, it's a lucky thing for the team and the crowd that I wasn't at the game and wasn't selected as the contestant.

I would have known the right thing to do. I would also have grabbed the microphone and launched into an explanation about the math behind the boxes.

The crowd was spared.

You, however, are not so lucky.

This is actually a famous mathematical problem called the 'Monty Hall Problem', after the US host of the old game show *Let's Make a Deal*.

While the origins of the problem itself go back to 1975, it exploded into national prominence when it was the subject of a column in *Parade* magazine written by Marilyn vos Savant on 9 September 1990. A reader sent in the problem and asked Savant for an answer. She replied, correctly, that you should always switch. The chances of you winning if you switch are ⅔, and only ⅓ if you stay with the same box.

A firestorm erupted. The column generated over 10,000 letters, the vast majority of which disagreed with her answer. Here are some excerpts:

> Since you seem to enjoy coming straight to the point, I'll do the same. In the following question and answer, you blew it. If one door is shown to be a loser, that information changes the probability of either remaining choice, neither of which has any reason to be more likely, to ½. As a professional mathematician, I'm very concerned with the general public's lack of mathematical skills. Please help by confessing your error, and in the future being more careful.

And

> Your answer to the question is an error. But if it is any consolation, many of my academic colleagues have been stumped by this problem.

So, if you agreed with my daughter and thought that switching wouldn't help, you were not alone. Savant followed up with another column in December, reiterating her answer that switching is better. After more correspondence, in February 1991, she called on school children to try it out for themselves and report back. She was universally vindicated as hordes of school children, at over a thousand schools, reported back that switching resulted in winning ⅔ of the time.

The affair peaked with a front-page article in *The New York Times* in July 1991. It has since spawned countless articles in mathematics journals, and has also touched philosophy, economics, and quantum mechanics.

But why is switching good?

Well, here's the way I like to explain it. In any probability problem, the most basic thing to do is to just list the possibilities.

In this case, there are three: the iPod must be in one of the three boxes. Let's say you pick the first box. There is a ⅓ chance the iPod is in your box and a ⅔ chance that the iPod is in another box.

Now the emcee opens a box containing the low-value prize and gives you the choice to switch. (This is an important detail – the emcee never reveals the box containing the iPod.)

Let's say the prize is in box one, the one you chose. In this case, switching is a losing move. If the prize started out in box two, then the emcee will open box three. In this case, switching is a winning move. And if it started out in box three, he will open box two, and switching will also be a winning move.

So, if you switch, you will win if the prize started out anywhere except the box you chose, which is a ⅔ chance. If you stick with your box, you will only win if that was the box that had the prize, which is a ⅓ chance.

Now, probability is hard. There are a lot of situations where intuition is wrong, and this is a classic example. If you're interested in learning more – a lot more – about the Monty Hall paradox, I highly recommend the book *The Monty Hall Problem* by Jason Rosenhouse.[6]

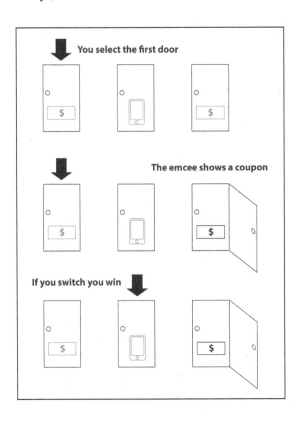

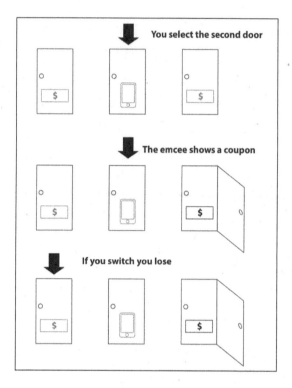

On average, in test after test, about 90 per cent of people stick with their first choice. Only 10 per cent of people switch. You would think that if people really believe there is a fifty-fifty chance, they might switch just for the fun of it, just for the variety. But they don't.

So, what's going on? Here we come back to two topics I covered earlier in this chapter: the endowment effect and loss aversion. People give more value to what they have in their possession. And losses make us feel much worse than gains make us feel good.

One of the things I love about games is the choices I have to make – I like to make decisions and see what impact they have on the course of the game and on my success (or

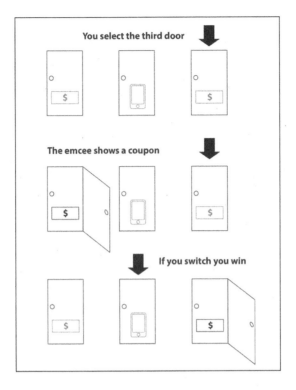

otherwise). And difficult choices make things that much more interesting.

But sometimes, of course — probably much more than I'm willing to admit — those decisions go wrong. And when one of my many decisions goes horribly wrong, I feel bad. I feel regret and remorse. And this is true for just about everyone.

Regret is normal and universal. And many studies have shown that people overwhelmingly would rather reduce the chance they feel regret than increase the chance they get a good outcome.

But the interesting part here is what happens when people change their minds. If you make a choice and it turns out to be wrong, you feel regret. If you make a choice and then change your mind and switch to another choice — and it turns out you should have stuck with your first choice — you feel *much more* regret.

I think we've all been in that position in school. If you go with a choice on a test and then later go back and change your answer, and it turns out you were right the first time — well, you feel a lot worse than if your first choice was wrong and you had stuck with it. In fact, although I have no evidence of this, I am convinced that the old adage that you should go with your first instinct is more of a reaction to not feeling bad rather than any improvement in your score.

How much more regret do you feel if you switch? It seems like it would be impossible to measure, but several experiments have tried. One involved the Monty Hall Problem itself. In this case, people were told that if they stayed with their original choice and won, they got a prize worth $10. But if they switched and won, the prize would be worth a multiple of that — people were told values ranging from twice as much to ten times as much. At triple the value for switching, about 50 per cent of people switched their choice. At ten times the value, about 90 per cent of people switched. So, this experiment puts the extra regret from second-guessing yourself at three times worse. Similar experiments have also produced results that show you feel about two to three times as bad when you change your mind and are wrong.

The gameplay applications of this are clear. One lesson is that if you don't want to feel bad, always go with your gut. But if you want to make better decisions, try to remember that you'll always want to stick with your initial choice, even if you come up with another idea afterwards. If you're aware of this tendency, you may be able to filter it out and hopefully make better choices – and maybe even win more often. Which – let's face it – pushes those regrets right out the window.

I'll leave you with another brain teaser. You're talking to a woman at a park and she says she has two children. Suddenly a boy runs up and grabs her hand and you ask if this is her son. She replies that he is.

What are the chances that her other child is a boy?

Most people say 50 per cent: the fact that one is a boy has no bearing on the gender of the other child. But this is not correct. There are four possible family arrangements that have two children: boy/boy, boy/girl, girl/boy and girl/girl. When I tell you that at least one is a boy, that eliminates one possible family: girl/girl. There are three combinations left: boy/boy, boy/girl, and girl/boy. Each is equally likely, but in only one out of three (and thus a ⅓ chance) is the other child a boy.

This is an example of what is called 'correlation', or in physics 'entanglement' (as in 'quantum entanglement'). Entanglement and correlation can cause all kind of interesting things to happen, but I will discuss that in Chapter 13.

Chapter 7

Who's the Best Judge?

Inherent Bias

You are in a store, holding a movie in your hand. You're planning to buy it, because you looked at the Internet Movie Database (IMDB) earlier, and saw it has a 7.5 out of 10 average rating from 200 people. But you haven't read any full-length reviews.

A stranger sees you standing there and comes over to you.

'I've watched that,' he says. 'It was awful. I didn't like it at all.' And he walks out of the store.

Would you still buy the movie?

Maybe you would, maybe you wouldn't. But studies have shown that most people would think pretty seriously about it before making a decision.

In other words, the impact of that one face-to-face discussion far outweighs the 200 opinions from online ratings aggregates.

But what has really changed? Before, there were 200 ratings with an average of 7.5. Now let's say the stranger in the store gives the movie a rating of 2. Now there are 201 opinions, with an average of 7.48.

Two-hundredths of a point really shouldn't matter. You don't know any more about the movie itself. The stranger's comments were very generic. So why does it make you think twice?

We are social creatures. As such, we are programmed to respond to individuals in a very different way than we do to abstract data. Our brain doesn't see this situation as 201 people reviewing something. It just sees two pieces of information:

- IMDB liked it.
- A stranger didn't like it.

Psychological studies have shown that we respond the least to numbers, more to written reviews, and most to face-to-face encounters

In one study, students were split into four groups, each given different information about courses, and asked to choose which they would take. One group was given just the course titles. A second group also got a numeric rating. The third group was given written reviews of the classes, although they were phrased in quite a generic way. And the fourth group had someone sit with them and verbally give them the same written review.

The numeric ratings had a slight impact on course selection, and the written review more. But the face-to-face encounter had the biggest impact by far — even though neither the written nor verbal presentations gave any basis for decision-making.

The study also found that more weight was given to negative reviews than to positive ones. One bad apple can indeed spoil the bunch.

You should be aware of this when you read a review – especially when you read one from someone you don't know. Of course, the more information there is in a review, the more you can get out of it. And the more reviews you read from a particular reviewer, the more you understand what they like and don't like, and how that matches up with your own preferences.

This is why more frequent reviewers can be more valuable than one-shot wonders, even if you don't always share their opinions.

So be aware that your brain is hardwired to give more weight to someone saying, 'I give this movie a 2, I didn't like it at all,' than just seeing a rating of 2 with no comment. Yet, in reality, they both have the same information. And 100 people giving a 7 has even more information, especially compared to the stranger making generic negative comments.

By understanding our inherent bias towards certain types of information, we can all make more informed decisions.

Prospect Theory

On a slight side note, but still within the realm of inherent bias in decision-making, I'd like to play a game with you. I'm going to give you two options.

Option A: You get $3000 guaranteed.

Option B: Roll a die. On a 1 to 5, you get $4000. On a 6 you get nothing.

Which would you pick?

Strictly mathematically, option B has a higher expectation value. Your average gain is $3200, and in option A it is $3000. But most people, about 80 per cent, choose option A, the sure thing.

Here's another game. I'm going to give you two new options.

A: You lose $3000 guaranteed.

B: Roll a die. On a 1 to 5 you lose $4000. On a 6, you lose nothing.

How about this one?

Well, in this case, a whopping 92 per cent of people take the second choice – even though it actually increases their expected loss.

In both examples, the vast majority of people will choose the worst option mathematically speaking. What's going on here? Is it just that people are bad at math?

This experiment was performed by two psychologists, Daniel Kahneman and Amos Tversky. They did a variety of experiments that repeatedly showed that people do not operate strictly by rational judgments. Based on their experiments, they developed a theory to explain this behaviour called 'prospect theory', which has developed into a fertile area of research, and offers a rich theoretical foundation for economics, finance, insurance, psychology and other areas. Kahneman received the Nobel Prize in 2002 for the development of prospect theory. Tversky would have shared the prize, but sadly he had passed away a few years earlier.

The amazing thing is that neither Kahneman nor Tversky was trained in economics. They were both cognitive psychologists and wrote many highly readable books about decision-making. I particularly recommend *Thinking, Fast and Slow*.[1]

Now, in a nutshell, prospect theory says a few things. First, people have different risk attitudes towards gains and losses: people are less likely to take risks to increase their gains, but they are more likely to take risks to avoid losses. This is clearly seen in the examples above, where most people wanted the sure $3000, but would take a gamble – a bad gamble, in this case – to avoid losing $3000.

Second, the way the problem is presented is critical. This is called 'framing'. You can present the same problem to people to make it seem like they are losing instead of gaining. So, you can manipulate what they decide.

For example, here's a slight variation on the original problem.

Participants were asked to imagine that the US was preparing for the outbreak of a disease that was expected to kill 600 people. The first group was asked to select between two different courses of action, with the following outcomes:

A: 200 people will be saved.

B: There is a ⅓ chance that 600 people will be saved, and a ⅔ chance that no people will be saved.

Over two-thirds (72 per cent) of people preferred option A, to definitely save 200 people.

A second group was presented with these outcomes:

C: 400 people will die.

D: There is a ⅓ chance that nobody will die, and a ⅔ chance that 600 people will die.

Now, in this framing, 78 per cent preferred program D, with the remaining 22 per cent opting for the program where 400 people die. However, A and C are exactly the same, as are B and D. The only difference is that A and B are expressed positively, in terms of saving people, versus in negative terms of people dying.

One game that explores this is *Deal or No Deal*. In this US TV game show, the contestant is presented with 26 briefcases, each of which has a hidden amount of money, ranging from $1 to $1,000,000. The contestant picks one to keep, and then selects cases to be opened, removing them from the game. So, the remaining pool of dollar amounts gradually shrinks.

At certain points in the game, the 'banker' makes the player an offer for their case, which they can either accept, ending the game, or reject in order to play on, opening more cases.

This game is basically a psychological exploration of prospect theory. A fair offer from the banker would be the average of the remaining dollar amounts. But the banker never offers this. They always offer significantly less. So, a strictly mathematically inclined person should never take the offer.

But then, a mathematical answer is not always a correct answer. Consider a simple game where you roll a die, doubling your money on a 2 to 5, and losing everything on a 6. Mathematically, you should continue to play *indefinitely*. Realistically, though, we know you have to stop at some point.

So, people do take the banker's deal. And the amount of money they are willing to leave on the table, to be able to walk away with a sure thing, helps economists fine-tune prospect theory, and learn more about people's risk tolerance.

Keep this in mind when negotiating in a game, or trying to anticipate what people what will do. Most people will take a sure gain, but take a risk to avoid a loss.

Review or Analysis?

When we talk about a board game 'critic', as with many things, the terminology can be a little fuzzy. Usually we are referring to somebody who critically reviews games, similar to a movie reviewer, or a film critic.

The second type of critic performs critical analysis — looking at themes, historical context and comparable works. Literary criticism is often of this ilk.

So, we've got two types of criticism: review and analysis.

These serve two completely different purposes, which leads to a lot of confusion.

Reviews are there to help you decide whether to buy a game, or see a film, or read a book. Analysis is there to help you think more deeply about the topic and see connections or features that weren't readily apparent at first glance. It can also help you understand the period in which the game was developed, or how the game reflects the designer, or how it fits into their larger body of work.

So, reviews by their very nature are time-sensitive. It is no accident that film reviews are typically published the day the film is released in theatres. Similarly, in the game world, it sometimes feels like there is a race between reviewers to get that first review up. It does seem that the earliest reviews garner the most attention, regardless of their ultimate utility. And it is rare that a review that comes out for a five-year-old product – whether it is a game or a film or a book – will attract that much attention.

Analysis is the exact opposite. Analysis requires time and perspective. Analysis requires that the critic be versed in many examples of whatever genre they are working in, be it poetry, literature, or film. Analysis requires expertise, and expertise requires time.

A game review does not require a game expert who has played hundreds or thousands of different games. Anyone can write a review that says what the game is about, how it plays, what works, what doesn't, and why they loved or hated it. A review is a snapshot of one person's reaction at a particular moment in time.

Sometimes reviewers attempt to incorporate analysis. *The New York Times Book Review*, for example, is famous for reviews that attempt to both review the book and place it in a larger literary context. But though I enjoy reading those reviews, I often find, at the end of a review, that I'm not sure whether I should buy the book or not. I'm just not given enough tools by the author, or enough personal bias, to make a judgment.

But there's a twist for games that doesn't exist for films, books, or even art. There is an expectation that board games will be played multiple times. There are certainly games that reveal all their depth on one playthrough, but others, especially classics like *Go, Chess* or *Bridge*, only reveal their depth and subtlety over time, and demand multiple plays.

So, are game reviewers obligated to play a game multiple times before rendering judgment?

This topic flared up in early 2012 over the game *A Few Acres of Snow*, where some felt that hidden depths or intricacies in the game were not being explored by early reviewers.

I disagree. I think there are certainly examples where those playing multiple times will have a deeper appreciation of the game, and that will come through in more in-depth reviews. And I think that the thoughtful reader of reviews will take that into consideration. Review is not analysis, however.

Analysis demands multiple plays. The ability to develop deep thoughts about a game – to put it into a context – requires familiarity with and exploration of the designer's intent. And regardless of the simplicity of the game, you can't get that with a single playthrough. If someone is doing an in-depth critical film analysis of *2001: A Space Odyssey*, or a dissection of *Hamlet*, I fully expect that person to have gone through the material more than one or two times. And I expect no less from game analysis.

So, as I read an article about a game, I try to put the article into the review or analysis bucket – and, depending on how it's classified, I approach it differently and have different expectations. From one, I am looking for quick reads on what a game is about and what works and what doesn't. For the other, I am looking for deeper analysis, multiple plays, and more thought. It is very rare for anyone to combine these two forms, for obvious reasons.

In the board game world, there are many reviewers, some better than others, but a real lack of analysis. And I think that this is the crux of the bemoaning of the state of board game criticism.[2] But I think it is unfair to lump these two together – review and analysis are really two different animals with two different expectations. If a review of *A Few Acres of Snow* doesn't discuss the presence of a killer strategy, that's fine by me. But a piece of critical analysis on the game would be sorely lacking if the killer strategy was omitted.

Chapter 8

Falling for Patterns

Quasicrystal Patterns

The 2011 Nobel Prize for Chemistry was awarded to Dan Shechtman of the Technion – the Israel Institute of Technology. He won it for his discovery, in 1982, of quasicrystals. What's a quasicrystal? I'm glad you asked.

The tale of the quasicrystal starts in the early 1960s, when mathematician Hao Wang was working on the problem of tiling a plane with square dominoes. 'Tiling a plane' (or 'tessellation') means completely filling a flat surface with shapes. We've known for a long time that there are only three 'regular' polygons that can be used to tile the plane: the triangle, the square, and the hexagon. You can make a grid of these shapes that extends to infinity. And that's why those shapes – squares and hexagons in particular – constantly show up on our game boards. It is very easy to drop those shapes on top of a board and have them regulate movement, or shooting, or whatever you'd like.

Actually, there are an infinite number of shapes that can be used to tile the plane that aren't regular polygons. The artist MC Escher was a huge fan of using shapes like lizards to tile planes in his drawings.

Tilings of the plane can also be accomplished by a collection of differently shaped tiles. Many bathroom floors are tiled in collections of different-sized square tiles, or squares and rectangles.

All of these tilings have one thing in common – they are periodic. This means that you can shift the whole pattern over and the tiles will line up with each other. This is called 'translational symmetry': you can translate the pattern onto itself and it repeats infinitely.

Mathematicians assumed that any tiling with a fixed number of tiles had to be periodic. There was no way to fill the plane with shapes and have the pattern never repeat.

Back to Hao Wang. In 1961, he was working on a problem with square dominoes with different colours on each side, and when the dominoes were placed down the coloured sides had to match up. He was basically taking an arbitrary set of these tiles and asking the question: 'Can I tell whether these tiles will let me completely tile the plane infinitely in all directions?'

He wasn't able to figure it out exactly. However, he was able to prove that if the tiling is periodic, you can always figure out if the set of tiles will in fact cover the plane. He thought all patterns must be periodic, so he figured this was as close to a proof as he was going to come.

Fast forward five years and mathematician Robert Berger was able to prove that you *can't* tell for sure if a set of Wang tiles will tile the plane. Therefore, using Wang's proof, there must be a set of square dominoes that fill the plane aperiodically — in other words, *without* repeating a pattern. Berger was able to find a set of 20,000 different tiles that would work. Unwieldy, but a breakthrough.

People kept working on aperiodic tiling, and in 1973 the brilliant physicist and mathematician Roger Penrose got it down to two tiles. With two specially shaped tiles, you can infinitely cover the plane, and not be able to shift the pattern onto itself.

These patterns typically display five-fold symmetry, and are fractal in nature. They look like they are repeating, but don't quite. They are very cool, and I've always wanted to buy Penrose tiles to cover my bathroom floor. Maybe someday.

Jumping forward in time to 1982, Shechtman was doing diffraction experiments on crystals to understand their shape and saw a five-fold symmetry pattern, which no one had ever seen before. A few more years of research convinced him that the crystal was actually aperiodic – unlike every other crystal seen to date. It was different all throughout the crystal. He named the new form of matter 'quasicrystals' and published his results, mostly to disbelief and ridicule. Chemist Linus Pauling said he was 'talking nonsense'. People were convinced there was no pattern in these crystals – only Shechtman could see there was. However, over the years, the existence of those quasicrystals was confirmed, and hundreds of others were discovered.

They have unusual properties, like high hardness, non-stick properties, and resistance to corrosion, that are still being explored.

'Hot Streaks' and other such Patterns

Humans are prone to what is called 'apophenia' on a larger scale too. This is the tendency for us to see patterns in random data, where there truly are no patterns. This is part of our evolutionary toolkit and while in general it helps us survive, it does have a bunch of side effects.

Superstitions can arise when certain disconnected events are joined in our minds because they occur together by the vagaries of chance. Notice that you're wearing the same underwear when your team wins a few games? Lucky undies are born.

This tendency is also very visible in casinos, where *Roulette* tables have giant poles displaying the last 10 or 20 numbers spun at that wheel, whether they were red or black, the percentage of low or high numbers, and a host of other statistics.

These things aren't cheap, I'm sure. Casinos spend a fair amount of money to buy and maintain them, plus it's extra work for the croupier to enter the data. So obviously *Roulette* players must prefer casinos that display these historical records.

I don't play *Roulette*. And I haven't talked to any *Roulette* players. But I've seen people scrutinise those lists like conspiracy theorists looking for the latest sign of Bigfoot. And you know what? If you look hard enough, you'll find something.

So, if these people walk up to a *Roulette* table and see that the last six numbers were black, what do they do with that information? Does that mean that red is more likely to come up next due to the 'law of averages'? Or does it mean that black is more likely to come up since the wheel is trending that way?

The answer – and this shouldn't be a surprise – is that the probability of red or black is exactly the same. The fact that there was a streak of a single colour in the last six spins is absolutely meaningless. The basic law of probability is that past results have absolutely no bearing on what will happen next.

Now, this goes against the grain for many people, so I'm going to say it again: the past history of independent random events has no bearing on the future.

A lot of people – including a lot of smart people – have a problem with this.

My father, who is intelligent and accomplished, plays *Craps*. Years ago, he shared with me his strategy.

'I wait for the shooter to get hot,' he confided in me. 'Then I bet big.'

Except there's no such thing as a hot streak. Every time you roll the pair of dice in *Craps*, you have a one in six chance of rolling a 7. Every time.

One of the reasons that humans have such an edge on the animal kingdom is that we're incredibly good at recognising patterns in chaotic data. This skill is arguably the foundation of intelligence. And while it has enabled us to develop all the science and technology that allows me to type this in New Jersey and send all the electrical signals and light waves scurrying around the globe to Australia, it can also get us into trouble.

And this is one of those ways.

Our brains are really good at perceiving a pattern in any set of data, even when there's no pattern to be seen. Sometimes randomness is truly random.

The Settlers of Catan is the grand-daddy of Euro games (games that generally emphasise trading and building, rather than eliminating the other players). Not only did it launch the Euro-game revolution, it's also a good game in its own right.

But there are those who don't like *Catan*, and the main complaint levied against it is also one of the reasons that it is such a good family game: the use of dice. Every round, players roll two six-sided dice and score resources based on the combined numbers rolled.

Players complain that there are always too many 4s, for example, or too few 9s, or an unbelievable number of 11s, whatever. And complaining about the dice rolls in *Catan* is like complaining about the price of popcorn at the movies — everybody does it, but it's part of the experience.

Now, one option for dealing with this 'luck' is to use a deck of cards instead of dice. In fact, there is a specific *Catan* product for this. This is a deck of 36 cards that have the standard distribution of the 36 possible dice combinations. So, if you go through the deck once, you'll always get exactly the expected quantity of each number. To add a little randomness, the bottom five cards are covered by a special card. When that is flipped, the deck is reshuffled.

In a four-player game of *Catan*, there are about 12 rounds, so 48 rolls in total. Typically, if you're playing with a dice deck you just reshuffle once – one and a half to two times through the deck is sufficient for most games. So, you're going to be very close to the expected distribution, especially for fringe numbers that have few cards in the deck.

But what are really the chances of getting extremely large or small numbers of these fringes? Let's look at 4s, 5s, 9s and 10s, since I hear the most complaints about those numbers being 'streaky'. Over the course of 48 rolls there should be four 4s, four 10s and a little over five each of 5s and 9s.

First off, I bet that's more than most people expect to see. I don't have any data on this, but I would speculate that people expect to see a 6, 7 or 8 on almost every roll. But, as I said, out of the 48 rolls, eight should be a 4 or 10 – one out of six. So, right off the bat, I think there are more of this type of number than people are expecting.

But let's dig a little deeper. As I said, out of 48 rolls in an average game there should be four 4s. But what are the chances of that in real life?

I ran a simulation of 1,000,000 games of *Catan*, and the results were interesting. Only 23 per cent of the time were there exactly the predicted number of 4s. In 20 per cent of cases, there were three 4s, and in 16 per cent of cases, there were five 4s. Two 4s and six 4s came up about 14 per cent and 10 per cent of the time respectively. So, the distribution curve of 4s over these 1,000,000 games is pretty flat and spread out. It's not a sharp peak at four rolls.

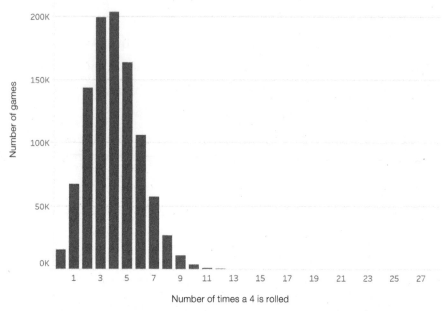

Number of 4s out of 48 dice rolls

Number of times a 4 is rolled

Let's say that having 50 per cent more or fewer rolls than expected is a 'streaky' result. In one-third of games this will happen for 4s, and there's also a one in three chance of 10s being 'streaky'. So, the chance of it happening for one or the other (or both) is fifty-fifty.

And this is just with those two numbers. If you add 2, 3, 5, 9, 11, and 12 into the mix, the chances of having 'streaky' results on some number during a game is quite high.

So, where dice give you a fairly flat distribution on the number of dice rolls expected, the dice deck gives you a sharp peak at the expected value. If you want to play with the dice deck, you can – but you have to understand that you are not playing the same game.

As an obvious example, the dice deck contains two 11s.

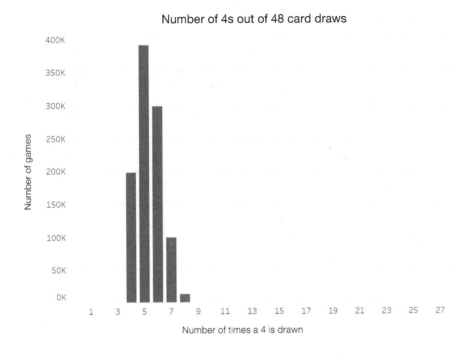

Number of 4s out of 48 card draws

If one of them comes up very early, it is much less attractive to take a chance on a future roll of 11. Conversely, if an 11 doesn't come up and you're halfway through the deck, the actions that reward you for rolling 11 start becoming really attractive.

If you're using dice, someone rolling or not rolling an 11 does not change the future probability of another 11 coming up. Just like with *Roulette* or *Craps*, the chance *looks* less probable, but is not. It's still a 4 per cent chance, no matter what happened in the game so far. But with a dice deck, that probability can go as high as 33 per cent.

In the end, I guess it's just a matter of taste. For me, the vagaries of the dice distribution in *Catan* and other games are just part of the game, and an enjoyable part at that.

The magicians Penn & Teller stage a magic show in Vegas. When I went to see them perform, Penn opened the show by saying: 'Welcome to Las Vegas – an entire city built on glorious bad math.'

Our brains see patterns everywhere. But we also have within us the ability to see what is real and what is a mirage.

There are no hot streaks. You're never 'due' for a roll. Just because you rolled low early doesn't mean you'll roll high later to make up for it.

Random is random.

Hindsight Bias

In 2012, just before the US presidential election was decided, I guaranteed on *The Dice Tower* podcast that one of two stories would be written:

> Mr Romney's victory was inevitable. The rocky economic climate was impossible for Mr Obama to overcome, and his modest success in foreign policy and other domestic programs were footnotes in an election year dominated by the economy.

> Mr Obama's victory was inevitable. From primaries through to the national elections, Mr Romney failed to connect with the average voter, and the power of incumbency was too large an advantage for him to overcome.

Sure enough, six months after Barack Obama was elected President, the outcome was considered inevitable and a foregone conclusion, and the pundits all agreed on the reasons why.

When we look back on a past event for which there was no consensus at the time, what happened is treated as the only thing that could have happened. We say, 'I knew it all along,' even though we really didn't.

In psychology, this is called 'hindsight bias', and it has been fairly well studied over the years. There are several factors that contribute to hindsight bias.

One is memory distortion: we simply don't remember what we thought at the time. Our knowledge is constantly being updated, and, once we know the results, we can emphasise certain details or even fabricate them to support the conclusion.

An experiment performed in 1999 by psychologist Linda Carli[1] asked two groups to read a story about a romantic encounter. The two stories were identical, except that one group's ended with a sexual assault and the other simply stopped before that outcome was revealed.

A week later, the groups were asked details about the story, and those who had read the one with the tragic ending emphasised and even fabricated details that they saw as 'warning signs' and made the man seem sinister, whereas the control group gave no indication that they even thought this was a possible outcome.

This is related to another tendency called 'sense-making'. Like the way our brains find patterns in the most abstract of places, they are also constantly trying to make sense of the world and understand what's going on and why. It turns out that the easier it is to come up with an explanation about why something happens, the more we think it is inevitable.

So, what's the problem with this? What's the disadvantage of thinking that an event was more predictable than it was? From an evolutionary standpoint, linking cause and effect, and therefore learning, was important for humans.

Well, there are several problems that hindsight bias can cause.

First, in many situations, we evaluate the performance of people based on the decisions they made with the available information. Hindsight bias makes it difficult to put yourself in the shoes of these people and objectively judge whether they made a justifiable decision or not.

In games, this type of second-guessing happens all the time. Whether it's a sporting event or a board game we're playing, afterward our choices are subject to criticism about why we made the decisions we made, and it's tough for people to put themselves in our shoes and understand the uncertainty we faced in the moment.

'Well, of course, if you raise there, your opponent will call your bet.' Well, maybe yes, maybe no.

Even when evaluating our own performance, it can be very difficult to recreate our mental state when we made the decision.

This even has legal ramifications. For example, in a medical malpractice suit, the decision that the doctor made is evaluated based on the information they had at the time — and that can be next to impossible to do.

The other issue is called 'myopia'. Far too often, when

we develop a reason why something happened, when we connect the dots, we stop looking at the event. So, if this reason is very easy to come up with, but wrong, then we are not learning the right lesson from the experience. In fact, a variety of experiments have clearly shown that the easier it is to come up with a reason why an outcome was inevitable, the larger the hindsight bias, and the greater the chance to make the wrong connection.

In games, this can be a big contributor to groupthink. After a game ends, there may be a facile explanation for why Suzanne won. 'Once she got this combo, it was inevitable she would win,' we may say.

The simpler the explanation is, the more likely we are to just accept that it really was an inevitable, unstoppable path to victory. The group will focus on that explanation and not look at the uncertainty that was present in the game as it was unfolding, and the other paths that were unexplored and untaken. And it can take a major shake-up in another direction before that inevitability is exploded as the myth it was.

So, how do we combat hindsight bias? Researchers suggest a few ideas.

First, simply being aware that hindsight bias exists can help you get past it.

Another technique is to examine causes for the outcome as close as possible to the event itself. The further away you are from the event before critically examining it, the more likely you are to accept a simple but flawed explanation and think the outcome was more inevitable than it was.

An experiment was performed with people watching a football game. One group was asked how inevitable the outcome was right at the conclusion of the game, and the second a few days later.[2]

The group that was asked immediately post-game thought that the outcome was substantially less certain than the group that was asked later.

So, whether you're playing a game or just navigating through life, pay attention to hindsight bias. It will make you more likely to examine your assumptions and get to the true root cause of what happened. Embracing the world in all its complexity is a challenging but ultimately worthwhile endeavour.

Power Creep

One of the complaints I hear about game expansions is that the new stuff is more powerful than the older stuff. It's called 'power creep' or, in the miniatures realm, 'codex creep', since in *Warhammer 40,000*, the most popular miniatures game, each army has a codex book. And when a new codex comes out, it seems like the new units are often the best. Cries of 'They just broke the game!' are shouted from the rooftops.

Many have a conspiracy theory about this. The game companies, they proclaim, need to sell expansions, whether they are cards or miniatures, to make money. If they don't make the new stuff better than the old stuff, people won't go out and buy it, and it will languish on the shelves. So, the game companies deliberately push this power creep upwards, making the expansions better and better, and forcing players onto an endless treadmill of purchases.

Well, I'm as much a fan of conspiracy theories as the next person (conspiracy theories, of course, being a great example of how we impose patterns on random data), but I think the explanation is perhaps more benign.

Let's take a hypothetical game and say it includes factions with a power level of five. That's just an arbitrary number, but higher power levels will be better than lower power levels.

Now let's say the company comes out with a new expansion that includes two new factions. We're going to give them the benefit of the doubt and say that they were trying to make both factions also have a power level of five and be balanced against all the existing factions.

But it is hard to make something exactly five. If you are making an interesting game, there are lots of different effects that will interact with each other, and make it tough to hit that precise power level. Just from natural factors, new factions are going to be distributed around five — some will be higher, some will be lower.

However, there's another element that comes into play. Playtesting is tough. And when you put an expansion in front of thousands of people, there is an excellent chance that they will figure out something that you missed. Almost always, players will find ways to make things work better and be more powerful than you anticipated. Combos will be found and exploited.

So, one of your new factions turns out to be a six, not a five. And players flock to it and enjoy beating up on all the other players. This is especially true in games that have an active tournament scene, like *Warhammer*, *Hearthstone* or *Magic: The Gathering*.

Okay, so you're a game company and the new expansion has turned out to be more powerful than the old stuff. But you truly want to have a balanced game system. What do you do?

You basically have two options. You can reduce the power of the new faction, or you can make the existing factions more powerful.

People like their stuff. Especially in miniatures games, where people spend a lot of money on the figures and a lot of time painting them, players become emotionally invested in their chosen faction and their units. They do not want to see their power reduced. Remember from Chapter 6, losing stuff is much more emotionally impactful than gaining it. People get so upset about this that they invented a word for it: 'nerfing'. Nerfing something means making it less powerful, less dangerous – trading in a real sword for a Nerf one. No one wants to get nerfed.

You hear about this all the time in video games. It is a lot easier for *World of Warcraft*, *Overwatch* or other online games to nerf a certain character by patching the game. It is much harder to do in a board game or card game. Those physical assets are out there in the real world.

I am aware of a few examples. In *Warmachine*, there was one warcaster – Lich Lord Asphyxious – who had such a powerful trick that publisher Privateer Press actually changed one of his abilities through an FAQ. The printed rules stayed the same, but tournament players were all aware of this change. In *Magic*, certain cards get banned.

Other than nerfing, the game company can make the other factions more powerful. And this is the technique that is used most often. The game company will add new units to existing factions, or change the rules. I maintain, with great confidence, that this is preferred by players to nerfing. Players would much rather get new toys to play with than have their old toys made less interesting.

So, the company tries to bring everyone up to power level 6. But, of course, some factions may accidentally get brought up to power level 7. And the ratchet keeps grinding on and on, jacking up the strengths and abilities in the game until it becomes unrecognisable from its original form.

This doesn't need to be some nefarious conspiracy. I never believe in a conspiracy when incompetence – or just plain missing the target – can explain the same thing.

Chapter 9

When 'Random' Isn't Random

Luck of the Dice

In the last chapter, I talked about how a random spread of dice rolls can sometimes lead our brains into falsely seeing a pattern. This leads to theories about 'hot streaks', *Roulette* or *Craps* 'strategies', and *Settlers of Catan* players seeing a disproportionate number of 4s.

But while humans excel at extracting order from chaos — and seeing patterns everywhere — we are lousy at creating randomness. We have definite way-too-specific ideas about what random looks like.

For example, if you're playing the *Powerball* lottery in the US, where you pick numbers between 1 and 69, there's only a small bit of strategy. The only thing you can really try to do to increase your potential winnings is to try to pick numbers that, if they come in, will make you the sole winner.

People pick birthdays and lucky numbers, but often they pick numbers that look random. Someone once asked me what numbers they should pick for a big $200,000,000 jackpot lottery, and I suggested 33, 34, 35, 36, 37, and 38 ... because I'm a wise guy.

They looked at me and said that I was crazy, and that there was no way that those numbers would come up. I tried to explain that it was just as likely as any other sequence – and if it did come up, the chances of someone else being crazy enough to pick them was pretty low, so they actually had a better chance of being the only winner and making more money.

But as they showed me their picks, they explained their nice random sample – some low, some high, but not equally spaced. They played with the numbers until they looked, to their eyes, like a good, carefully crafted random sequence.

Now, I tell this story not to poke fun at anyone, but to emphasise that this is a battle we all have to fight against our instincts.

Another instinct we have is to assume that rolling *more* dice in a game will result in the game being *more* random.

Luck in games is like salt. Some people never want it, and rely on things like dice decks to combat it, while others pour it on. And based on their preferences, gamers may look at a game and say, 'Too much luck for me!' But what does 'too much luck' really mean?

Let's take a look at so-called 'bucket-of-dice' games. These are games like *Warhammer* where you throw 10, 20, 30 dice at a time. If you like lots of luck, these are the games for you, right? Lots of dice means lots of luck.

Well, not really. Counterintuitively, more dice means *less* luck.

The definition of more or less luck in this case is the number of times you get a result that is outside what you expect. Let's say I'm playing a shooting game where I roll a six-sided die and on a 4, 5 or 6, I score a hit. If I roll four dice, the average would be two hits. My expectation value is hitting on 50 per cent of the dice. If I get lucky, I get extra hits. Let's say a lucky throw is where I get 75 per cent hits – three hits out of four dice.

How often do 75 per cent or more of my dice hit?

With four dice, three hits will happen 31 per cent of the time.

Now let's double up and roll eight dice. Everyone knows that my expectation is four hits. But the chance that 75 per cent of the dice will hit (that's six or more hits) is no longer 31 per cent. It is 14 per cent.

If I roll 16 dice, the chances of rolling 75 per cent hits drops to only 4 per cent. The more dice you roll, the greater the chance that the final result will be tightly clustered around the expectation value and hence less luck.

When you roll several dice for hits, like in these examples, the results can be represented by the familiar bell curve. Most of the results will be toward the peak in the centre, and results that are out on the tails will happen less frequently.

As you roll more dice, the curve gets narrower and sharper around the expected value. It becomes more and more likely that the random result will be close to the average. This is why an apparently even spread of dice rolls across a game of *Catan* can feel wildly random, despite only using two dice.

1 Die

2 Dice

3 Dice

4 Dice

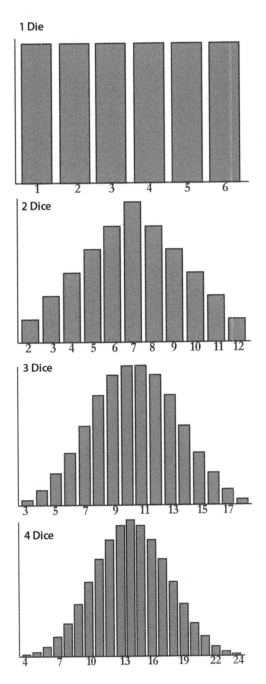

The peak gets sharper and sharper as more dice are rolled, so there's an increased chance the number rolled will be close to the average.

Few people have an intuitive feel for this effect. Several scientific experiments have been performed to highlight it. Say I ask people, 'Which is least likely: three or more heads out of four coin flips, six or more heads out of eight coin flips, or twelve or more heads out of sixteen coin flips?' Almost everyone will say the chances are the same. And even those who correctly say 12 heads out of 16 is less likely will invariably underestimate just how much less likely it really is. Remember the difference between 3 hits on four dice and 12 hits on 16 dice is 31 per cent versus 4 per cent.

Wargames, which simulate battles and campaigns, are known for being highly strategic, and rewarding skilful play. Luck plays a role, but the victory goes to the better general.

Let's compare two strategic wargames about World War II.

Rise and Decline of the Third Reich is an old wargame from publisher Avalon Hill and is one of the driest games you will ever play. Cardboard counters represent armies, and move across a hexagonal grid. Combat is resolved by a single die roll, and, at a few points during the game, the difference between a 1 and a 6 can be huge.

On the other side of our comparison, we have *Europe Engulfed*, a wooden block game from publisher GMT Games. This is as close to a Euro strategic World War II game as you'll find. Major battles will easily see 30 or more dice tossed, probably more than are thrown in an entire game of *Third Reich*. This is a classic bucket-of-dice game.

And yet it has less luck than *Third Reich*. With 30 dice, you can pretty much rely on the expected value being the result. The chance of hitting on 75 per cent of your dice: 0.2 per cent.

So, remember, more dice equals less luck. When you're playing *Twilight Imperium*, *Warhammer*, *Europe Engulfed*, *Axis and Allies* or any other game where lots of dice get tossed and someone turns up their nose at you for playing a 'luck-based game', you know what to say.

Pseudorandomness

Gamer Scott Nesin has a seven-foot-tall dice tower (a box filled with slanted platforms that randomises dice dropped into it). He built it to generate random numbers for his website, GamesByEmail, instead of using a computer to do it. Well, that's his stated reason, but I'm sure he just built it because he thought it would be cool.

Anyway, he said that one of his motivations for using 'live' dice instead of computer-generated random numbers is that dice were more random. This is interesting – what exactly does 'more random' mean? And is he correct in saying that his dice tower would be 'more random' than computer-generated numbers?

Random number generators can be placed into two categories: hardware-based and software-based. Hardware-based random number generators rely on some physical phenomenon to generate the random numbers. The Pentium chip from Intel, for example, uses thermal noise in the processor to generate random numbers. This is a kind of 'white noise' (like an old TV set to no particular channel –for more on white noise, see chapter 15), where each individual measure of frequency is completely unconnected with the frequency at the next point. In other words, it jumps around randomly.

Other things that have been used include disk drive seek times, time between keystrokes, and time between network packet arrivals. Many online *Poker* sites rely on combinations of these factors to shuffle their decks. And of course, dice, cards and *Roulette* wheels count as physical hardware random number generators as well.

The problem with using physical phenomena is that it is possible for the determined hacker to either figure it out or manipulate it in some way to bias the output. If you can monitor what is happening in the system, you may be able to use that to determine at least part of an encryption key.

Also, it is difficult to determine how 'random' these phenomena actually are. Maybe time between keystrokes varies in predictable ways depending on what people are doing. Maybe thermal noise fluctuations have a particular pattern in particular chips. It is almost impossible to know if there are hidden biases in these random sequences.

The other type of random number generator is based on software algorithms. These start with a seed number and manipulate that to figure out the next number in the sequence. This is called a 'pseudorandom' number generator, because, given the seed and the algorithm, you can figure out the next number in the sequence. So, it's not truly random in the classic sense – it is a deterministic sequence of numbers.

Randomised shuffling is actually a big deal for online casinos. There were a few instances in the early days of online *Poker* where people were able to crack the random shuffle.

Shuffling algorithms on a computer start with a random seed. They then proceed deterministically to shuffle based on the seed. So, two shuffles with the same seed come out the same. The problem with early shuffling systems was that they used 32-bit seeds, which give about 4 billion possible shuffles (out of a possible 8×10^{67} ways of arranging a deck – that's an 8 with 67 zeroes after it!). So, you can only end up with a very, *very* small fraction of the possible number of decks. You would never see over 99 per cent of possible deck configurations.

To compound the problem, some early online *Poker* systems based their seed on the system clock. So, hackers could narrow the possible seeds down to about 200,000, and then it was within the capability of the computer to quickly figure out those 200,000 decks and match them up with the cards that the player could see – in *Texas Hold 'em*, that's your hole cards and the community cards. Then, once you know exactly what the deck is, you know exactly what everyone has and is going to have.

Current online *Poker* sites have much better seed generators. You need at least 226 bits to uniquely identify all possible decks. So, they use 256-bit seeds. They also randomise based on time between player actions, not the system clock, making it much, much harder to hack.

The first work on pseudorandom number generators was done by computer pioneer John von Neumann, back in the 1940s. Random numbers are a critical part of computer science, powering networking, encryption and security.

Mathematician Robert Coveyou, who also did pioneering work on pseudorandom numbers, famously said, 'Random number generation is too important to be left to chance.'

It's a lot easier to test pseudorandom number generators to see how random they are than natural random processes like radioactive decay, so they are subject to a lot of analysis. There are a battery of tests that have been designed to see how close these sequences come to a theoretically ideal random sequence.

Work on pseudorandom number generators continues to this day. The current state-of-the-art generator is an algorithm that was developed in the late 1990s called the Mersenne Twister, which is very fast and generates random numbers of very high quality.

Before you decide that all this talk of the quality of pseudorandom numbers is just pie-in-the-sky theory, let me tell you a quick story about a game with bad numbers.

Back in the 1990s, a computer programmer named Ronald Harris worked for the Nevada Gaming Commission. His job was to check the source software code for slot machines and other gambling devices. While reviewing the code for a new *Keno* machine (*Keno* is kind of like a lottery, where you are picking numbers from a grid), he realised that there was a flaw in the pseudorandom number generator it used, and that it would repeat sequences in a predictable way. He found a 'trigger sequence', and once he saw that sequence he knew how the next few games would end up. Rather than report his findings, he decided to try to cash in on them.

He and an accomplice went to Bally's casino in Atlantic City, where he knew the machine had just been installed. He watched the numbers from his hotel room and, when he saw his trigger sequence, he called the accomplice and told him to go and buy the winning ticket for the next game.

Sure enough, the numbers came up as predicted and the ticket hit for over $100,000, which is unheard of in *Keno*.

Unfortunately for Mr Harris, his accomplice did not have ID, which is required by New Jersey law to collect a prize that large, and authorities became suspicious. The scheme quickly unravelled and Mr Harris ended up in jail for a few years.

See? Bad random numbers can matter!

There are flaws in pseudorandom number generators, even in the Mersenne Twister. And, as noted earlier, hardware-based random numbers, like those generated by a seven-foot-tall dice tower, can also have statistical biases. The biases in both pseudorandom and hardware-based random numbers are so small, though, that, without looking at tens of thousands of numbers with high-speed computers, you would be hard-pressed to find the flaws. I know that many believe that they are victimised by online dice rollers, but, like with the *Catan* dice rolls, that's just a difference between the reality of what *is* random and what people *perceive* is random.

So, while seven-foot dice towers look really cool, they won't help you be any more random.

Disorder in a Deck of Cards

If you shuffle a deck of cards and then deal five out on the table in front of you, sometimes they will fall into a significant arrangement. For instance, let's say you are trying to draw a recognised *Poker* hand. Well, the five cards you pull out may be more or less significant depending on the game you are playing, and someone not versed in the rules might just see them as 'random'.

But what makes one arrangement more 'random' than another one?

In the late 1800s, a Viennese student by the name of Ludwig Boltzmann was pondering this question. He discovered a paper that had been recently written by the incredibly insightful mathematician and physicist James Clerk Maxwell, which determined the statistical distribution of the speeds of particles in a gas. As the temperature of a gas increases, the particles move faster. But they won't all move at the same speed. They are all zipping around and bouncing into each other, which speeds up some particles and slows down others. By looking at the math behind this, Maxwell was able to determine, for a given temperature, volume and pressure, what percentage of particles had different speeds.

The fundamental insight that Boltzmann had was to flip this description upside down. Rather than start with the big features of the gas – like temperature or pressure – he started with what the particles were doing, and figured out what that tells us about the temperature or pressure.

Let's look at a box of gas molecules. What Boltzmann realised was this: every configuration of particles in the box is perfectly legal, just like every configuration of cards in a deck. There's no law that says that every single particle can't be on the left side of the box, or that every Spade can't fall on the top of the deck. Random chance can result in that happening. If we measure the pressure of the gas at any one moment, we might find that it is high on the left side and low on the right. If we measure the variation of Spades versus other suits in the deck, we might find that it is high on the top and low on the bottom.

Each particular arrangement of particles or cards is called a microstate. And each microstate will give a different measurement of the pressure on the box or distribution in the deck.

But the vast majority of microstates will give the *same* reading – one that shows the particles or cards spread out relatively evenly throughout the box. The more microstates there are that result in a particular distribution, the more likely it is that you will find the box or deck in that state. Vastly, *vastly* more likely, as it turns out, as there is an astronomically huge number of states that give an even distribution. A balloon, for example, contains something like 10^{22} molecules – that's 10 with 22 zeroes. It is basically a certainty that we will find a balloon in one of the even-pressure states. Even a deck of 52 cards will result in a massive variety of distribution states – most of which we would consider 'random'.

*

To go back to something more tangible, in the previous chapter I showed how rolling 16 dice versus 4 dice can lead to a massive increase in your ability to predict the distribution of 'hits' (rolling a 4, 5 or 6).

Let's take it to the extreme and say I have thousands of six-sided dice and I roll all of them. Yes, it is possible that they all hit. But it is incredibly likely that they will be really, really close to half hits and half misses. If I have 10 trillion trillion dice? I'm not doing the math, but it is going to be close to fifty-fifty to a very high degree of accuracy.

Boltzmann determined that the measure of disorder in a system (called 'entropy') is equal to the 'logarithm' of the number of microstates that lead to the arrangement you are looking for. A logarithm is conceptually one less than the number of digits in a number. So, 1 has one digit and so its logarithm is 0, 10 has two digits and its logarithm is 1, the logarithm of 100 is 2, and so on.

Taking this to the dice example, there is only one way things can be ordered so that every die scores a hit — so the entropy is 0. But, with just a hundred dice, there are 10^{30} ways to get exactly 50 hits and 50 misses. That's a really big number — 30 digits long — so the entropy would be around 29.

Remember, higher entropy is, in a sense, more disorder. What Boltzmann showed is that to measure disorder, you need to be measuring something big about the system. In our gas example, we are measuring the amount of pressure in the box. In our dice example, we are measuring the number of hits. The disorder depends on what you're measuring. In our *Poker* example at the beginning of this section, we were measuring the chances of drawing a recognisable *Poker* hand. Since those are actually quite common, the disorder in the deck may be very orderly indeed.

Similarly, Boltzmann showed that the Second Law of Thermodynamics (which says that entropy in a system always increases) isn't really a law at all. If you keep shuffling a deck of cards, it may come back to the original arrangement. If you wait long enough, all the gas molecules in a box may end up on the left side. The 'law' in this case is a statement of probability. Yes, it *may* happen. But for most systems, you will have to wait the duration of the entire universe to see it. In practical terms, the law *always* holds.

Chapter 10

Games and Entropy

Entropy

How many riffle shuffles are required to randomise a deck? Back in 1992, mathematician Persi Diaconis of Stanford University, along with Dave Bayer of Columbia University, figured out that the answer is about seven.[1]

Most *Blackjack* dealers were doing significantly fewer, which caused consternation in Vegas. However, in a later study in 2011, Diaconis then determined that for games where suit is not important, like *Blackjack*, four shuffles will suffice.[2] So, the casinos are safe.

Of course, one of the keys to shuffling is that you don't do it perfectly. A perfect riffle shuffle is one where you split the deck exactly in half and precisely interleave the cards. With enough practice, it is possible to be able to do this repeatedly. And 52 consecutive perfect riffle shuffles will return a deck of playing cards back to its original configuration. So, with a perfect shuffle, you know the state of the deck. You are looking for an imperfect shuffle – which is my specialty!

*

Shuffling is a classic example of one of the key elements of thermodynamic science, which I brought up in the previous chapter: entropy. Remember, entropy is a measure of the randomness or disorder of a system. In any system, the Second Law of Thermodynamics says, entropy will increase over time. For example, a nicely organised game, all sorted and bagged, has low entropy. If you just throw all the pieces in the box and shake, you have high entropy.

Entropy has an interesting place in the history of science. All of Newton's laws, which govern the movement and interaction of particles, and even all the laws of quantum mechanics, are time-reversible. They work the same whether you move forward or backward in time. Yet entropy seemed to give time a definite direction – the increase in entropy defines the arrow of time.

As we saw, this mystery was resolved in the late 1800s by Boltzmann, with his development of statistical mechanics. If you have a box of 1000 pennies and you shake it, there are many, many more combinations where there will be about 500 heads and 500 tails than that there will be 1000 heads and zero tails. The state where there are equal heads and tails is a high-entropy state, while 'all heads' or 'all tails' is a very low-entropy (or highly ordered) state, and it requires energy (sorting through and flipping the tails to heads) to go back from the high- to low-entropy state.

As mentioned previously, Boltzmann formulated that it is possible, but extremely unlikely, that a system will move from a disordered state to a more ordered state. But if you reverse all the forces on the box, the system will return to its ordered state, like billiard balls scattered over the table reforming into their starting triangle shape.

You Too Can Be Unique

There is a tremendously large number of ways a deck of cards can be ordered. So one day I was thinking – if I shuffle a deck, what are the chances that there has never been another shuffled deck exactly like that in the history of humankind? Can I make a deck that has never been seen before in the world?

Well, we need two pieces of information to figure this out. The first is: how many ways are there to arrange a deck of 52 cards? This one is easy to figure out, even if it is a big number. It's $52 \times 51 \times 50 \times 49 \ldots$ all the way down to 1. This has a short name called '52 factorial' (or 'factorial 52'), and it's written like this: '52!'.

For example, '4 factorial' or $4! = 4 \times 3 \times 2 \times 1 = 24$.

Factorials get really big, really fast. So big that 52! is about 8×10^{67} – or 8 with 67 zeroes after it. That's a staggeringly large number. The number of atoms in the observable universe is estimated to be about 10^{80}, which is not that much bigger in the grand scheme of things – only about a trillion times bigger.

The other piece of information we need is: how many shuffles have occurred in the history of humans? Now, we can't exactly determine this, so we need to fall back on our friend the estimate. Some people never see a deck of cards in their lives. Some people sit at a *Blackjack* table and see lots of shuffled decks during a day. I'd like to err on what I think is the high side, so let's say that on average, one deck of cards is shuffled per person on Earth per day. So, if there are 7 billion people on Earth, that's 7 billion shuffles per day, or 2.5 trillion shuffles per year.

But that's just now, when there are 7 billion people. The modern deck of cards goes back to about 1500 AD, when there were less than a billion people. So, we need to figure out how many total person-years there have been, and then convert that to days.

Because I am insane, I got a variety of statistics about the estimated world population from year to year. Adding everything together, there have been about 850 billion person-years from 1500 until 2012. Multiply by 365 and you get a tad over 300 trillion person-days. So that's 300 trillion shuffles since the invention of the card deck. I really think this is too high, considering it includes every single man, woman and child on all continents, since the year 1500, but let's go with 300 trillion for now.

300 trillion is expressed in scientific notation as 3×10^{14}. That's 3 with 14 zeroes after it. Remember that our deck of cards can be arranged in 8×10^{67} ways. So, the total number of shuffles that have occurred in history is a really, really small fraction of the total number of possible shuffles. If you shuffle a deck right now, the chances it ends up the same as one of those 300 trillion shuffles is 1 in 10^{53} — that's 0.0000000000 ... 43 more zeroes ... and then 1. Very, very small.

So, Dr Mulcahy's assertion that 'you can be unique – just keep shuffling and you'll get there' is not quite right. A single shuffle should be more than enough.

Let's say you have just opened a new deck of cards, ready to shuffle your own unique deck. If you take a look at the cards, you will see that all the suits are together and each goes from ace to king in order.

Now give the deck a single riffle shuffle. There are still clumps of cards of the same suit, in order.

Give it another riffle shuffle.

When you started, you had a deck that was highly ordered. Now you have gradually made the deck more and more disordered.

We discussed earlier, in Chapter 9, how the disorder in a system is roughly equivalent to the scientific concept of entropy. As you shuffle the deck, the entropy of it increases, until it reaches a state of maximum entropy, where it will stay.

Entropy is a key concept in the branch of science called thermodynamics, which is the study of heat and energy. The 'discovery' of entropy goes back to the work of the French mathematician Lazare Carnot, who was studying steam engines. This was right at the time when steam engines were beginning to fuel the Industrial Revolution, and Carnot was interested in how an ideal steam engine would perform if it were perfectly designed – if there were no friction, no leakage, just a perfect system. He found that, even if everything were mechanically perfect, the amount of energy that you could extract from the engine would decrease over time.

In the 1850s, German scientist Rudolf Clausius expanded this concept, and maintained that there was a fundamental quantity which represented, in essence, energy that could no longer be used to perform work. He called this 'entropy'. Ultimately, this led to the Second Law of Thermodynamics, which (if you remember) says that the entropy of a system will *always* increase over time. As you may also recall, Ludwig Boltzmann later used statistical mechanics to show that this is not so much a law as a statistical rule – but we still call it a law.

This 'law' is frequently misunderstood. Very often we see order in the world, and we see areas where it is increasing, not decreasing. But the Second Law looks at the entire system, not just part of it.

For example, you can restore order to your deck of cards. But you need to expend energy to sort the cards and put them where you want them – for instance, by playing the game Solitaire. And to do that, you are increasing the entropy in your body and the world around you. You can decrease the entropy in your body – and in fact, we all must in order to stay alive. Life is a very ordered state, with low entropy. So, we constantly have to fight against the increase in entropy. We do that by taking energy into our bodies – energy that almost invariably comes from the Sun. The Sun is a huge entropy sink. The entropy of the Sun is constantly increasing, allowing us to lower our entropy here on Earth.

Entropy is a weird scientific concept. It is not something that you can capture or put in a bottle. In the 1700s, scientists thought that heat was represented by a fluid called 'caloric'. But by the mid-1800s, it was clear that something else was going on.

During that time, there were two big questions around this, which had no answer. Boltzmann answered the first, determining that the entropy in a system is equal to the logarithm of the number of microstates that lead to the arrangement you are looking for. A manufactured deck of cards comes in only one arrangement, and so has an entropy of zero (1 microstate $= \log 1 = 0$). A shuffled deck of cards comes in numerous arrangements and has a much higher entropy.

The other question goes back to Isaac Newton and his theories of motion from the 1600s. Everything in those equations shows that they are exactly time-reversible.

So, if I can just play the movie of me shuffling the deck of cards backward, I have actually decreased entropy, right? There's nothing physically impossible about doing that.

The answer to that paradox is that it is true — if you set everything up just right, and time everything perfectly, you can 'rewind the movie' in real life. But it would need to be an incredibly specific setup, the forces would need to be precisely calibrated and timed, and any heat or other energy that was lost by the system needs to be introduced at exactly the right time and place. In other words, it is incredibly unlikely to just happen on its own, and requires a huge amount of effort on our part to set it up.

Entropy and Board Games

So, what does this have to do with games? Almost every board game is a series of states. If we are measuring something in a game – let's say the score – we can calculate what the entropy is based on those states.

In most games – I would suggest that in all good games – the number of states where you win is much smaller than the total number of states. It is a more ordered state than most. So, from a statistical mechanics standpoint, we are moving from a high-entropy state – disorder – to a low-entropy state – order.

Game mechanics have to move the game from a high- to low-entropy state. In a strategy game, the decisions of the players will affect the game state. In a luck-based game, there may be random elements that serve that purpose.

However, there are several entropy trajectories that a game system can characteristically take. In some games, the entropy – or disorder – is the highest at the beginning of the game. They start in a totally random state and gradually order emerges.

Many card games fit into this category. In *Hearts*, a classic trick-taking game, the disorder is highest at the start of the hand. As it progresses, fewer cards are out, more information is learned about the card distribution, and entropy drops as the hand comes to a close.

Other games have an entropy arc. They start in a very ordered state and end in a very ordered state. But in between, disorder grows. *Chess* is a good example of this type of game. The game always starts in a known state, and the early moves are highly scripted. But as pieces are moved out of their starting positions and have more options, the number of game states explodes. Then, as pieces are traded away, the number of game states is reduced and entropy drops. Of course, not every *Chess* game reaches this state. Sometimes they end in a high entropy state. But ending in a lower entropy state is typical. Personally, games with this low-high-low arc are my favourites.

Then there are games where the entropy is basically flat throughout. The disorder or complexity of the system is pretty much the same. Basketball is a good example here, and I'm sure you can think of many others.

There are other elements we can tease out by looking at games through this lens. In a physical system, in order to counteract the increase of entropy, the tendency to drift into the most common situation, you need to expend energy. Similarly, in a game, if we want to shift the game state into a lower entropy state, we usually need to expend some resource. Typically, this is an in-game resource, like money or stone or bricks. But it can also simply be mental energy to see how to manipulate the game system to get it where we want it – and eliminate the possibility of entering undesirable states.

This way of thinking about games – the number of possible states, and how those grow and shrink through the natural course of the game and the actions of the players – gives both designers and players another tool for interpreting the flow of action and energy throughout the game.

Big Numbers and the Origin of Life

If you roll 10 trillion trillion dice, there is a very good chance that 50 per cent of them will be 'hits' (rolling a 4, 5 or 6). The percentage will vary slightly each time you roll the dice, but the math of big numbers says that it will be very consistent – not exact, but consistent. Similarly, a single deck of cards has 52! (or 8×10^{67}) possible microstates.

The fact that combining small numbers of things can give you a crazy number of combinations is really important in science. Combining just 52 cards gave us almost as many combinations for a deck as there are atoms in the observable universe.

For example, one of the theories of the origin of life is based on this concept. Here's the idea.

Say you have a chemical reaction that turns molecule A into molecule B. But there are certain molecules that help turn A into B much faster. These are called catalysts. Our bodies are filled with catalysts, mostly proteins, that help turn things into other things faster than they would do spontaneously.

Now, if I have a reaction that turns A into B, and I just throw a random molecule C in there, it is very unlikely that C will act as a catalyst. Maybe it's one in a million. But if I have lots of different molecules, the number of groups increases really quickly.

In a 52-card deck, there are over 2500 different ordered pairs. If I have a deck of 1000 different cards, I have 1,000,000 different pairs. So now, if something has a one-in-a-million chance of being a catalyst, there's a very good chance that in a thousand molecules there will be one that's the right catalyst.

In places like undersea volcanic vents, where a whole bunch of different molecules are forced into the same area, the number of different combinations of molecules grows very quickly. Eventually, the number of combinations will be much bigger than the chance that there is a molecule that will be a catalyst for two others.

And just like with the deck of cards, ultimately you may get a combination that loops back onto itself. So, C helps turn A into B, and D helps turn B into C, and A helps turn D into E. The picture below tries to illustrate this, as a chain of reactions loops back onto itself, and the molecules help make more and more of each other.

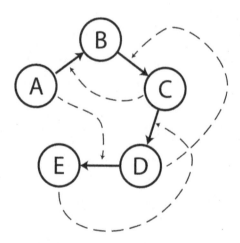

An unrealistically simple autocatalytic set. The solid lines show transitions between molecules. The dashed lines show which molecule helps the transition. so C helps A turn into B, and A helps D turn into E.

This is what's called an 'autocatalytic set', and the theory says that once you get an autocatalytic set, you have ignition on the engine of life. And while the chances of a specific set of molecules being autocatalytic is really small, the number of such sets grows incredibly quickly with each additional type of molecule you have, the same way that adding more and more cards to a deck makes the number of possible arrangements incredibly large. Eventually, the number of combinations of molecules gets so big that it becomes statistically impossible that an autocatalytic loop *won't* develop.

Who knew that the key to life's origins might lie in a deck of cards?[3]

Chapter 11

The Evolution of Life

On the Origin of Species

2019 marks the 210th birthday of Charles Darwin, and the 160th anniversary of the publication of his (rather lengthily named) book *On the Origin of Species by Means of Natural Selection, or The Preservation of Favoured Races in the Struggle for Life*. Darwin's insights, while they have been modified and extended over the years, form the foundation of biological theory.

Let's start with a quick and oversimplified overview of evolutionary theory. First, evolution requires random variation within a population – variations that can be passed down to the next generation. For example, some individuals in a population may be faster than the average. Maybe their legs are structured slightly differently, or they have a better means of storing and releasing energy.

What the theory of evolution says is that variations that lead to more offspring in the next generation will lead to that trait making up a larger percentage of the population.

That's it. Traits that are better at reproducing themselves will become more widespread.

Of course, there's a huge body of work around this, but that's the core of it. So back to our faster animal. If the herd is constantly being chased by predators and slower animals get caught, they won't have as much of an opportunity to reproduce as the faster animal. Now, of course there's a random element at play – a faster critter may trip over a rock and get caught anyway – but on average, over long time spans and many generations, the traits that reproduce better will predominate.

The flip side of random variation in the creature is random variation in the environment. If it's getting colder, then creatures that have traits that help them in the cold may reproduce better. The key thing is that these traits aren't developed *in response to* the change in the environment – they were there all along and just became more important and more helpful as the environment shifted.

So, with this background in mind, let's look at some games with evolutionary themes.

First up is *Quirks*, published in 1980 by Eon Products. It was designed by the same team as *Cosmic Encounter* and *Dune*. In *Quirks*, you control plants and animals that are made up of various cards. For example, you can put an elephant head with a turtle-shell body and porcupine tail. A lot of the fun is just putting the creatures together and seeing what they look like.

The environment is represented by a marker that moves down a track, so the world changes from jungle to plains to desert and back again. Each trait card has a different value depending on the environment. On your turn, you can change your plant or animal and then try to 'attack' another creature to gain dominance.

Quirks is a fun game, but probably the weakest from the Eon team. It's very fiddly, with lots of looking stuff up. It does, however, earn the distinction of being the only game I'm aware of where you play a plant.

Next up is *Tyranno Ex* from Karl-Heinz Schmiel, designer of the great *Die Macher*. It was published in 1992 by Moskito, and later republished by Avalon Hill.

Tyranno Ex focuses on dinosaurs. You can play dinosaur cards onto the track, and each is rated for which environment it likes. However, unlike other evolution-themed games, the players cannot change the creature traits — the play centres around modifying the environments. There are four tracks, and players try to manipulate the environment on each track. Dinosaurs that match more of the environment gain a bonus when they attack one another.

Tyranno Ex is an enjoyable game that has fallen by the wayside. If you can find it, I suggest picking it up and giving it a shot. The way you modify the environment instead of the creature is an interesting twist.

Next, we move ahead to 1997 for the Doris & Frank game *Primordial Soup*, published by Z-Man Games. In this game, you play a group of amoebas floating around in a pond. Your amoebas have to eat cubes of the other players' colours, and then poop out cubes of your own colour, which other players can then eat. Each turn, players earn points that they can use to either reproduce, help their amoebas move, or buy gene cards, which give special abilities – like being able to eat different types of food, carry food with you, or attack other players' amoebas.

The changing environment is represented by the way the distribution of food cubes changes over time. This gives a more organic feel to environment change than the abstract tracks from *Quirks* or *Tyranno Ex*.

Primordial Soup is a lot of fun, and always feels to me like a 'tactical' evolution game. You are down in the trenches with your amoebas, worrying about what they'll eat, how much damage they have, where they can put babies, etc.

Finally, I'd like to discuss *Evo* by Philippe Keyaerts, designer of *Vinci* and *Small World*. *Evo* takes us back to the dinosaurs. You control a small herd of critters that starts off on the beach of an island. You have to move, reproduce and survive to get as many of your dinosaurs on the board as possible, since that's how you score victory points each turn.

Each player has a playing mat showing what their dinosaur currently looks like. On each turn, several attributes come up for auction, like extra legs, horns or fur. These give your dinosaur improved abilities to survive. The twist is that you have to relinquish victory points in order to gain genes, so gauging how much a gene is worth is a key part of the game.

As the environment changes in *Evo*, different spaces on the board become hospitable to all the dinosaurs, so it forces them to move around to stay alive, giving the game a dynamic feel. When first learning to play, it's best with three players as it can drag. But once you understand the system, *Evo* moves along at a nice clip, with some fun decision-making along the way.

A few similar games include *Urland*, *Evolution*, *American Megafauna* and *Wildlife*. All are interesting and worth checking out.

However, as you can tell, none of these games really does a good job of modelling evolutionary theory, because, well, it just doesn't make a fun game. A good evolutionary theory game would inherently have no decision-making; evolution is just random variation taking advantage of the current environment. But as long as you don't come away believing that amoebas purchased their tentacles, you are on the right track.

Groupthink

One of my favourite games is *Through the Ages*. It is a multifaceted game simulating the development of civilisation from the Stone Age to modern times, with the goal of collecting the most 'culture' as possible. There are many interrelated systems. You need to build mines to gain resources to build structures like libraries, temples and wonders; build farms to feed your people and grow your population; and establish a military to defend against aggression or, if you are so inclined, just take what you want from your opponents. You can also gain leaders who give your civilisation special abilities.

I enjoy reading discussions about the game on BoardGameGeek, especially when people talk about winning strategies, seemingly without ever coming to any kind of consensus. Some people love lots of farms, some like military, others like libraries. Some say that whoever gets Michelangelo as a leader will win; others swear by Leonardo da Vinci or Napoleon.

Interestingly, when people post about these topics, it's invariably couched with the phrase, 'In my group, everyone thinks that X is the way to go.' Or they describe the last three games they played, and why one strategy was dominant.

I've played the game dozens of times now, and I don't think I'm any closer to figuring out what the 'best strategy' is. I bring this up, however, not to discuss *Through the Ages*, but to talk about the phenomenon of 'groupthink' in gaming.

In this context, groupthink occurs when a group of players play together most of the time, and their play tends to settle on a single strategy as the best. The term itself, though, implies that the strategy is not actually the best one.

One way to understand groupthink is to use a metaphor from the world of biology. But to get there, we will first have to go through our old friends math and science.

Imagine that all the possible strategies for *Through the Ages* are represented as bumps and pits spread out across a large mattress. The deeper a pit is on the mattress, the better that strategy is for the game. New players are like balls rolling around on the surface. They may initially try a strategy that they are familiar with in a similar game, or they may try some random actions to see what happens.

Eventually, they will roll into a pit. But their pit may not be the deepest pit on the sheet. In mathematics, this is called a 'local minimum', as opposed to the best strategy, which is called the 'global minimum'. Other players will end up in different pits, and in the end the other players will see which one is better. They will then tend to migrate their strategies towards that player's minimum.

In physics and chemistry, this type of surface is often used to visualise the energy states of a system. When a chemical compound is in a minimum, it is in a stable state. In order to move a molecule, or the ball, out of that minimum and into another minimum, energy is required. And the deeper the pit, the more energy is required.

I think this metaphor applies to games as well. To move out of a particular strategy and to a different place on the landscape requires energy from the player. And whether the player can exert enough energy to move to the next minimum depends on the player and the game.

Certain games have good strategies that are relatively close together on our strategy surface. In other games, they may be further apart, meaning that they require more radical departures from existing strategies rather than small adjustments.

Or they may be longer, more complex games. In shorter games, it's less of an energy investment to try out radically different strategies. But in longer games, like *Through the Ages*, the investment in time to play makes players less likely to try. It's like setting out across a desert in the hopes of finding an oasis in the distance.

Getting back to our physics metaphor – long, complex multi-strategy games require much more energy to explore the landscape. And so, groupthink is born, as players optimise around the particular minimum they are in.

Most games are more complex than the energy surface metaphor from physics. They are actually more akin to what is, in biology, called a 'fitness landscape'. We've got the same mattress, except this time the local minima are creatures that are more 'fit' from a reproductive standpoint. (Going back to the theory of evolution, these are creatures that are more suited to their environment and have a better chance at reproducing.) A population randomly spreads out over a portion of the fitness landscape, through recombination, when genes get shuffled between parents, and mutation, where new genes emerge from random errors in copying. And that's the way it explores the fitness landscape and evolves into more fit creatures.

Fitness landscapes are more complex than energy surfaces for a variety of reasons. But the reason we're interested in here is that fitness landscapes change based on changes in the other creatures in the environment. If cheetahs evolve to become faster, the strategies that prey were using, like running swiftly, may no longer be enough to save them. In game terms, the actions and strategies of the other players affect what the strategy surface looks like for you. In community-reliant games, such as *Magic: The Gathering* this is often called the game's 'meta'.

If all the players are in a particular local minimum, it may start to make other strategies more or less valuable. So, the play style of certain groups not only presents an energy barrier to changing strategies, but may also turn what are good strategies in other groups into bad strategies in yours.

In biology, a small group breeding among themselves will typically lose fitness over time, as harmful mutations gradually accumulate. Similarly, to get out of a gaming rut requires either a trailblazing explorer willing to risk loss to try new things or, more typically, the cross-fertilisation of play styles among a variety of groups.

That's why playing at a convention or in a tournament, or even online, can be so refreshing. If a game is to have longevity and long-term strategic interest, in my opinion, it should have a balance of strategies. If one strategy is overused, another should become more attractive.

I suggest that if a game has a single global minimum – a best strategy, in other words – it will grow stale after a time. Navigating a constantly changing strategy surface leads to games of endless fascination and depth.

Ratings and Regression toward the Mean

We play board games for a variety of reasons – social interaction, intellectual challenge, exploring innovative mechanics, or learning about a new topic. But let's face it – all of us, at least once in a while, like to win.

So, we play, and sometimes we win, and sometimes we lose. Now, let's say you sit down to play a game and you have one of those magical performances. You understand exactly what's going on. You can predict what your opponents are going to do. Your strategies unfold precisely according to plan.

We've all been there. We've all been in that zone.

But then you play the next time, and, unlike your previous commanding performance, you just can't seem to get ahead of the curve. You're back in the middle of the pack.

Well, this happens a lot. At least to me.

So, what's going on?

Nobel Prize-winning psychologist Daniel Kahneman tells an illuminating story in his 1982 book, *Judgment Under Uncertainty*. While attending a Navy conference, he spoke with naval flight instructors about whether they thought that praise or criticism was a better motivating tool. And almost universally they picked criticism.

One instructor put it this way: 'If a student has a good landing and I praise them, they usually do worse next time. But if they do poorly and I yell? They always improve on their next landing.'

This is a classic example of what is called 'regression toward the mean'. In a nutshell, if you do poorly one time, you will most likely do better the next. And if you do well, you will most likely do poorly next.

Assume that your performance at an activity – whether it's landing airplanes, bowling or playing *Primordial Soup* – is shaped like a bell curve. Most of the time, you have an average performance, with a lesser chance of being better or worse. Note that this average is for the way you personally perform. If your average at ten-pin bowling is 120, sometimes you may bowl a 140, and sometimes a 100. But obviously someone else will have a different average.

Now, in our bowling example, if you bowl that 140, you may feel pretty good about yourself. You may think that your skills are finally improving and that you're 'in the groove'. However, from a purely statistical standpoint, it is much more likely that your average is still around 120. So, when you bowl again, there is much more of a chance of bowling lower than 140 than higher. Now, this can work both ways: if you bowl lousy, you'll be more likely to do better next time.

In both cases, you will tend to drift back (or 'regress') toward your average (or 'mean'). Hence the term 'regression toward the mean'. But this isn't the only factor at play when it comes to determining skill level.

In 1999, two psychologists from Cornell University, David Dunning and Justin Kruger, did a series of studies to measure how skilled people thought they were.[1] They gave people tests in three different areas: logic, grammar and humour. (For the humour test, the experimenters collected a series of jokes, and first asked a panel of professional stand-up comics to rank them on a scale of 1 to 11. They then used the average for each joke to determine what the 'correct' answer should be.)

Then a bunch of Cornell students took the tests. Afterward, the researchers asked them to estimate how they had done on the tests relative to their peers, as well as what percentage of questions they thought they had gotten right.

Regardless of the type of test and regardless of age or gender, or any other variables, the results were the same. People who did poorly on the tests (we will call them the 'unskilled' here) on average ranked themselves much, much higher than their actual performance. On the humour test, for example, the people who performed among the worst – at the 12th percentile level – thought that they had performed at the 58th percentile level.

It's like the old joke: most drivers think they are above average. (Obviously half of people are below average, but no one thinks of themselves that way.)

On the other hand, people who did well on the tests – those that scored around the 90th percentile – thought they had performed around the 70th percentile.

This effect became known as the Dunning–Krueger effect. People that are unskilled at a task may know they're not the best, but they dramatically underestimate how unskilled they really are. And those at the top, who realise how much they know and don't know, tend to underestimate just how skilled they are.

Interestingly, studies have also shown that once someone begins to get trained in an area – once they become just a little bit more skilled – their estimates of how good they are get much closer to reality. They start to see just how much more there is to learn, and begin to see the outlines of everything they will need to know to truly become skilled at the task. In the beginning, however, our lack of knowledge makes us susceptible to the illusion that we are closer to being skilled than we are.

I think about this effect often in the gaming world. We live in an age where many of us flit from game to game. So few of us are actually experts in any given game. We may be very good at gaming in general, but rare is the game that we will play five or more times.

Which is why we must rely on analysis and not casual review before accepting that a game is 'broken' or that there isn't much strategy – particularly with a more complex title like *A Few Acres of Snow*. It has been shown that Dunning–Krueger-like effects are much more prevalent in complex topics. Maybe a strategy was dominant in your first play of a game, but that doesn't mean there aren't counter-measures. It doesn't mean that each game will play out the same way.

Now, admittedly, sometimes it does. And there are games that I *would* classify as broken. But most of the time, designers have played their games probably hundreds of times. They are truly experts. And it is the height of hubris to assume that we, in our first play of a game, see something obvious that the designer has missed.

As a personal example, in one of my designs, *Space Cadets: Dice Duel,* you roll lots of dice. Some folks have posted on BoardGameGeek that, after a single play, it's obvious that the game is all luck. Whoever rolls fastest, or rolls the best symbols will win the game. There's no point to playing.

This drives me nuts! As the game's designer, I've played it literally hundreds of times, and there is a ton of subtle stuff going on: gambits and counter-gambits that experienced crews can use on each other, and layers to all the tactics. I just want to reach out and challenge these people to a game against me and my kids and see how long they last.

On the flip side, when we released the game, I was concerned I wasn't expert *enough*. I was worried that there was some stratagem or tactic I had missed that would indeed blow the game wide open. In other words, I was on the top end of the Dunning–Krueger situation. I was really an expert – probably *the* biggest expert in the world (except for my daughter) – and I was concerned that I wasn't as good as I thought, that there was something out there I could learn to make me better.

We are all susceptible to the Dunning–Krueger effect, both for things we are skilled and unskilled at. This is not about being 'stupid' or 'incompetent'. It's a natural human flaw we all share. But by being aware of it, we can recognise that we are very poor judges of our own skill.

So how *can* you tell if you are really improving, or if it's just a statistical fluke? Well, that's a tough question. And one that people spend a lot of time working on.

As a starting point, it is reasonable to assume that it is easier to improve at an activity when you first start doing it. Almost always, your biggest improvement in how you play a game is between your first and second games. You have a clearer idea of the goals, techniques and tricks that may bring victory. But after you play something 10 or 20 times, it's harder to improve. The areas where you can improve are smaller and the lessons you can learn have only marginal impact.

In addition, the level of luck in a game also influences how much you can trust whether you are actually improving or just getting lucky. Games with more luck will have a broader, flatter bell curve. So, it is more likely that your performance will fall away from the mean. And a pure luck game will have a flat bell curve. Your performance can range anywhere from first to last place with no ability to affect the outcome.

However, even pure strategy games, like *Chess*, will have some spread over possible performances. We are human, so mood, temperament, how much sleep we got last night, and whether we just drank a jumbo Slurpee can all impact our performances. So, when trying to evaluate if we are improving or not, there will always be some uncertainty in the short term.

This has implications for organising bodies like the United States Chess Federation (USCF). They spend a lot of time developing and fine-tuning their rating systems to be able to predict what the average performance will be while taking into account possible random variation of people's play.

The USCF ratings are on a scale from zero to 3000. The math behind the ratings assumes that your actual performance in a game will be in a bell-curve distribution centred around your rating (that is, that you will regress towards your own personal mean). If your rating is 1700, you may play one game at an 1800 level, and the next at a 1650 level. The assumption is that the player who has the highest-rating performance for a particular game will win that game.

Think about it like this: imagine that each player is carrying around a box with different pieces of paper inside. Each slip of paper has a number on it that represents the strength of their game. If my rating is 1800, my box will have a lot of slips with numbers near 1800, some further away from 1800, and maybe one really high and one really low. When I play another player, instead of actually playing the game, we each reach into our box and pull out a slip. Whoever pulls out a higher number wins the game.

If I play someone who has a rating of 1900, there is more of a chance that they will pull a higher number out of their box than I will – but there's still a chance that I could win. The standard deviation of a chess rating is estimated at 100 points, just based on the vagaries of day-to-day performance.

Based on the difference between the ratings, there is an expected win value. If the ratings differ by 200 points, it is expected that the higher-rated player will win 75 per cent of the time. For a 300-point difference, that jumps up to an 85 per cent chance. These expected win values are used to update people's ratings.

The formula for updating ratings is pretty straightforward. If you're in a tournament, you add up your expected win values for all your games and compare that with how you actually did. You then multiply the difference by a factor – called 'K' in the USCF system – and add that to your rating. If you perform worse than expected, your rating goes down; if you perform better, it goes up. This K-factor decreases as you play more games. So early on, your rating is more volatile, and as you play more games, it decreases. And over time it is harder for your rating to shift.

There are many interesting features to this rating system. First, it is completely relative. If two players are both rated 1500 and play each other 50 times, and evenly split the games, they will both end up being rated 1500, since they performed at expectation. But most likely they both will have improved over those games. So if there are isolated communities of chess players, their ratings can get out of sync with other groups. This is similar to the idea of 'groupthink' we talked about in the last section. Inter-group play is critical for this rating system to work well.

Since it is relative, the pool of ratings as a whole is subject to drift, as players come in and leave. In particular, new, improving players typically should be rated higher than they are – the rating system has a tough time keeping up with them and tends to lag behind. This has the effect of causing 'rating deflation'. Over time, people's ratings at the top levels will drift downward.

Now since this is a relative system, it shouldn't matter – since they are drifting together. But people don't treat it that way. Even though the USCF tells people not to obsess about the absolute value of their rating, at the same time they declare that players reach Master level when they get a rating of 2200, and Grandmaster at 2400. So, they kind of undercut their own argument.

Ideally, someone rated 2200 today should have a 50 per cent chance of beating someone rated 2200 fifty years ago – but that is impossible to say. Chess theory and the overall level of competition are changing, which changes the absolute meaning of a particular rating. There have been some suggestions to try to set a formal benchmark that would help stabilise the ratings. One idea is to define a standard chess-playing computer that would be entered in tournaments periodically. Since its strength cannot change unless the software is changed, it would represent an absolute level of strength that could be used to anchor the system. But people play differently against computers in tournament settings, and it is possible that players will learn tricks to use against the computer. So, this idea has not gained any traction.

Since players tend to get upset when their ratings go down (our old friend loss aversion again, from Chapter 6), especially when it's through no fault of their own, the USCF has added a 'ratings floor' to their system (determined by taking your rating, subtracting 200 points, and rounding down to the nearest multiple of 100). If your rating is 1743 your rating floor would be 1500. Your rating can never, in your entire life, drop below your rating floor. While it helps people's egos, it further distorts the rating system and moves it away from its intended goal.

In an attempt to sidestep a lot of the deficiencies with the current rating system, the USCF has implemented a 'title' system. This is similar to an achievement system in a video game. If, for example, you win five tournaments while rated higher than 1800, you get a particular title. Unlike rating points, you can never lose a title – you just accumulate them. Sort of like my game collection.

So, the bottom line is that even the USCF will freely admit that the ratings as absolute numbers are meaningless. The basic purpose of a rating system is to estimate what the chances are of each player winning a game. And it does a pretty good job at that.[2]

Chapter 12

What's Good about Being 'Good'?

Size and Fairness

Games are often used by researchers in the social sciences to explore human behaviour. One of the largest global studies was published in 2006,[1] and the results were pretty interesting. The study was performed by a team from institutions around the world, including Caltech, Oxford University, the University of British Columbia, Universidad de Los Andes in Bogotá, Colombia, and many more. Thousands of people in a wide variety of diverse populations all played the same three games to look at how they interact with strangers. The participants included people in the jungles of New Guinea, nomadic tribes in Kenya, fishermen in Siberia, and workers in Missouri.

The three games that were played were pretty simple. They involved a stake of money, which was set at a day's pay in that particular region, but which for simplicity we will set at $100. The participants got to keep the money after the experiment. Each game was a one-shot interaction with another person, and was anonymous. They didn't see each other at all, and they knew they would never play the game against that person again.

The first was the *Dictator Game.* In this game, one player is given the stake and they can divide it any way they want with the other player, who must take the money. In this case, from a strictly mathematical standpoint, player one should offer nothing and keep the whole stake. The offer to player two is a measure of the fairness of player one.

The second is called the *Ultimatum Game.* Again, one player gets all the money and decides how much, if anything, to give to the other player. Before the game starts, however, player two secretly sets a level below which they will refuse the offer, in which case both players get nothing. So, if player two sets the level at $20 and the offer is split 85/15, neither player gets anything. But if the offer is 70/30, player one will get $70, and player two will get $30. In the case of the *Ultimatum Game*, again, going strictly by the 'Nash Equilibrium', player two should accept any offer of $1 or above. Anything is better than nothing.

As we discussed in Chapter 1, the Nash Equilibrium is a set of choices that gives all players their best result. Not all games have what is called a 'fixed' Nash Equilibrium, where the player does the same move every time – *Rock-Paper-Scissors* is a trivial example – but the *Ultimatum Game* does. You should always offer $1.

In reality, of course, when scientists conduct this experiment, the rejection rate of a $1 offer is basically 100 per cent. No one accepts it. In fact, very few people ever *offer* that little – under 1 per cent of players. The amounts offered most frequently are between $30 and $50.

For children, though, the results show that kindergarteners are more rational. They'll make more asymmetric offers, and accept them more frequently. This gradually shifts as we age, until it stabilises at typical adult values around age 20.

Scientists have interpreted the results of the *Ultimatum Game* to mean that there is an inherent 'fairness doctrine' in people that enforces certain social norms, even at the expense of their own wellbeing. However, it doesn't take much to upset that apple cart.

For instance, try a three-player variation of the *Ultimatum Game*. Player one still makes a single offer with two parts: what they keep and what's up for grabs. But players two and three each secretly either accept or reject the offer. If they both reject, nobody (including player one) gets anything. If only one of them accepts the offer, that player gets the offered money. If they both accept, the offered money is randomly given to either player two or player three.

The introduction of this element of competition drastically changes the results of the game. In experiments, the typical offer drops from $40 to $20. And the $20 offer, which was rejected 80 per cent of the time in the two-player version, is now accepted 70 per cent of the time.

If it's increased to five players competing to get the offer, the mean offer drops to $10, still accepted by someone 70 per cent of the time.

So, competition trumps fairness. Which I guess is not all that surprising.

The third game in the study also brought in a third player, but to punish rather than to compete. This is called the *Third-party Punishment Game*. In this game, player one gets a stake, and decides how much to give to player two, who has no choice but to take it. Player three starts the game with an amount of money equal to half a stake. Player three can set a minimum level at which they will punish player one if their offer to player two is at that amount or below. If player three punishes, player three loses 20 per cent of the money they are given, and player one loses 30 per cent of the initial stake.

So, let's say player one gets $100 and offers $10 to player two, keeping $90. Player three starts with $50, half of the $100. If player three decides to punish, they will lose $10 (20 per cent of $50), and player one will lose $30 (30 per cent of the initial $100). Player one ends up getting $60, player two ends up with $10, and player three ends up with $40. The numbers aren't important – the main point is that player three gains nothing by punishing. If they don't punish at all, they just walk away with $50. They have to pay to anonymously punish what they consider an unfair offer.

The hypothesis of the researchers was that people from societies that are more integrated – that have more markets, and therefore more economic and social interactions with strangers – would tend to give higher offers. They called this factor 'market integration' and used the percentage of food that was purchased in a market, as opposed to gathered or hunted, as a proxy for this value. And the offer value did indeed rise very noticeably as the market integration increased. In the *Dictator Game*, the Hadza tribe of Tanzania had the lowest offer, at an average of 26 per cent. The citizens of the Midwest US state Missouri had the highest average offer, at 47 per cent. The other groups were in between those values, with a very strong correlation between increasing market integration and increasing *Dictator Game* offers.

Surprisingly, the other parameters that were examined – including income, wealth and household size – showed no correlation with offer size. It was purchasing food at the market that gave the greatest indication of offer size. Similar results were seen in offers in the *Ultimatum Game* and the *Third-party Punishment Game*.

Interestingly, there was no difference in the average offers for the *Dictator Game* and the *Ultimatum Game*. The *Third-party Punishment Game*, though, showed a significant drop in the average offer. The threat of someone not involved in the transaction giving up something to punish the giver is apparently not taken very seriously. This was an unexpected finding, and is an interesting area for future research.

On the punishment side, the biggest correlation with how willing people were to punish was with the size of their community. Regardless of wealth, market integration, or income levels, people from larger communities were much more willing to punish – sacrificing their own wellbeing to do so – than people from smaller communities. People from a community of fewer than 100 people barely punished at all; those from communities of 5000 or more punished, on average, any offer below 45 per cent of the stake. That's a huge difference.

The researchers hypothesised that in small communities, the reputation of individuals is more important than anonymous punishment, and acts as a corrective factor in future transactions in daily life. However, in large communities, reputation is often impossible to determine, and anonymous punishment on the spot is required as part of the culture.

I'm not sure if that's correct, but something is going on.

One of my favourite games implements a form of the *Ultimatum Game*. *Junta* is a tongue-in-cheek simulation of power politics in a banana republic.

At the start of each turn, one of the players, 'El Presidente', draws a random amount of money as the budget – but the total amount is unknown to the other players. El Presidente then proposes how the money will be divvied up, and the players vote on it.

If the budget passes, the money is distributed. If not, all the money is kept by El Presidente – which sounds good on paper, but can actually be bad, as having a large stack inevitably leads to coups and assassinations.

So, games are not just for fun. Used correctly, they can be a potent tool in the arsenal of social scientists. This is a rich area for research that continues to be mined.

Cheaters Do Not Call Themselves Cheaters

There are always competing interests between individuals and groups, and these conflicts lead to cheating. For example, everyone in a town may agree to pay some money to fund a police department to keep the town safe. Let's call that contribution 'taxes'. If one or two people don't pay their taxes, the police department still has enough money and protects the town – including those who didn't pay. From the point of view of those who don't pay their taxes, they come out ahead – they get the benefit of police, but have extra money in their pockets. Self-interest trumps group interest. But if too many people do this, the system collapses and no one gets police, and everyone ends up worse off. As a society, therefore, we have an interest in stopping cheating – whether that society is our country, co-workers or game group.

This whole 'cheating versus cooperating' issue goes back to our old friend from Chapter 3, the Prisoner's Dilemma. But the bottom line is that situations like this tend to end up with everyone cheating (that is, defecting) – unless there is something that changes the balance. And that something is establishing a cost for cheating. Something that pushes the Prisoner's Dilemma in a different direction where self-interest doesn't trump group interest (that is, where there is an incentive to cooperate).

Over thousands of years, human society has come up with ways to deal with this, techniques that in a very real sense have allowed us to develop culture as we know it today. Perhaps the most powerful of these is reputation.

In my game *Space Cadets*, one of the minigames requires that the player in charge of 'sensors' finds a particularly shaped piece in a bag by using touch alone, before a 30-second timer runs out. In an early demonstration of the game, a player pulled the correct piece out of the bag a second after the timer had expired. This was a pivotal moment in the game. It was really important that the players gain the target lock that would result from the success of the sensors player. After pulling the piece out, he looked at me with a questioning look on his face – would I allow it?

Now, I wasn't playing in this game. I was simply teaching, and answering questions. So, I tossed it back to the players. I said that if they wanted to count it as good, they should count it as good. Which they did. They got that critical sensor lock and, honestly, I don't even recall if they won the game or not.

This incident got me thinking about cheating and its partner, trust. *Space Cadets* is purely cooperative. So, while the players can cheat, who is really the victim? If the players manipulate what happens so that they have fun, is that wrong?

Me? I don't think it's a problem. I treat *Space Cadets* as a toolkit for creating a fun experience, and if the players want to fine-tune something or take a do-over or whatever – it's their game and they should be allowed to do what they want. Of course, if they do that and then complain that it's too easy – well, that's not the designer's problem.

This also comes up in semi-cooperative games like *Descent* or *Dungeons & Dragons*. The Overlord or Dungeon Master player is the adversary of all the other players. But there are many Overlords who don't always go for the throat, who play not to win at all costs, but to make it fun for the adventurers. For some people, playing at less than full tilt is tantamount to cheating, but this will vary from group to group. Again, this is not something that harms another player.

There are, of course, cheating behaviours that *do* harm other players.

Way back — close to 25 years ago now — I was at a games convention playing *1830*, a railroad game. I had been working on an intricate plan for several turns, the culmination of which required me to purchase certain shares of stock. As the time came, I realised that I was literally $1 short. It wasn't like I needed $5 and had $4. I needed $214 and I had $213. I was really, really close. Without that one extra dollar, my plan would disintegrate and the game would effectively be over for me.

So, I cheated. I threw my stack of ones onto the bank's pile of ones, and they got mixed in before anyone could count how much I had placed. One player accused me of 'splashing the pot' — a term from *Poker* where you announce your bet and toss your chips into the middle, making them hard to count. I apologised and said I wouldn't do it again. The other player was suspicious but let it go.

Honestly (and I guess I'm using that word ironically now), I don't even remember if I won the game or not. That instance of cheating on my part is burned into my memory, and not in a good way. I know that this makes me look bad, and maybe this will give some of you second thoughts about playing a game with me in the future. I hope not. I will say that this is not the only time that I've cheated in a game, but the number is very small, and it hasn't happened in a long time.

I tell this story to illustrate a few things about cheating.

First, research has consistently shown that people who cheat do not consider themselves bad people in the moment.[2] They may feel guilt afterward, but at the time it seems like a perfectly reasonable and justified action.

In my case, I felt that the game and the situation were unfair to me. I had a perfectly good plan. I had saved up money ($213 worth) and being short a measly $1 was destroying the entire thing. The difference between having $213 and $214 was huge in game terms, but really, really minor in terms of money. That $1 represented a huge swing for me and that just wasn't fair.

Research shows that few people will cheat someone out of a lot, but many people will cheat a little bit. In experiments where people are given the opportunity to cheat, a sizable percentage (around 30 per cent) will cheat a little. A much smaller percentage (less than 1 per cent) will cheat a lot. In my case, I considered this a little bit of cheating. I splashed the pile of ones; I am quite sure I would not have splashed the pile of hundreds.

I agonised over whether to tell this example. It is a perfect illustration of why people cheat, and personal stories always make things more interesting. But I know that by telling that story I am sacrificing some of my reputation as an honest and trustworthy guy. I worried about this, even though this happened so long ago, and is way overshadowed (I hope) by the rest of my gaming life.

Preservation of our reputation, and concepts of 'fairness' are two really important (often competing) psychological motivators for us, as the studies on the *Dictator Game*, *Ultimatum Game* and *Third-party Punishment Game* should show. But studies have also shown that cheaters almost always give the justification of a sense of something being unfair, or something being owed to us, and that cheating is a way of rectifying that situation.

Rules are at the heart of the gaming community. So, arguably, cheating is of special importance to us, as opposed to society at large. Most of us simply assume that people won't cheat, and won't play with people who have a reputation for doing so. We have built up many structures – not just in gaming, but in society as a whole – to deter and detect cheating.

But let me leave you with this. Most of us believe that the vast majority of gamers don't cheat when we play a game. But if that's the case, why, when playing a blind-bidding game like *Fist of Dragonstones*, do we make players put their bid in their fist? Why not just have each player decide and then announce what their bid is?

Chapter 13

Tic-Tac-Toe and Entangled Pairs

Tic-Tac-Tanglement

Let's play a game. It's a very old game, tracing its origins to Ancient Egypt and the Roman Empire. It is also a game that has been 'solved' by use of a Nash Equilibrium – two experienced players playing their best response will always end in a tie. The game is *Tic-Tac-Toe*.

In *Tic-Tac-Toe* (or *Three Men's Morris* or *Noughts and Crosses*), players take turns drawing their symbol (usually an X or an O) on a three-by-three grid. The first player to get three symbols in a row, diagonally or orthogonally, wins. (This is similar to a known math problem, but I will come back to that later.)

Why am I talking about *Tic-Tac-Toe*? Because, believe it or not, it is a gateway game to gaining a more intuitive feel for quantum mechanics, and entanglement in particular.

In 2006, Allan Goff published an article in the *American Journal of Physics* entitled 'Quantum Tic-Tac-Toe: A Teaching

Metaphor for Superposition in Quantum Mechanics'.[1] As you can tell from the title, he specifically designed the game to help people wrap their heads around entanglement. At the same time, he invented a pretty nifty game mechanic that might be usable in the 'real world' with a little tuning.

Here's the way it works. You play on a regular *Tic-Tac-Toe* board, with the regular objective. But when you move, you get to make two marks on the board in two different spaces, and you tag them with the move number.

So, for example, on their first turn, the player X will put X1 in two different spaces. Then player O will put O2 in two different spaces – including spaces that include the original X marks. Then X goes again, and marks two spaces with X3, and so on. (X always marks with odd numbers, and O with even numbers.)

These spaces do not have actual Xs or Os in them yet – not yet. They just have 'virtual' Xs and Os. When you place two X1 marks you are saying that one of those will turn out to be the real X, and the other will be erased. But you don't know which is which when you place them – the two spaces are 'entangled'.

Eventually, as multiple marks are placed in the same space, you will form a loop of connected spaces. It's easy to see once you look at a board example. Basically, if you can form a chain of marks – going from one X1 to an O2 to an X3 and then back to X1, for example – then you have what is called an 'observation'. At this point, you must resolve the chain of spaces to decide what actually is in each space: X or O? The player who did *not* play the last mark gets to decide.

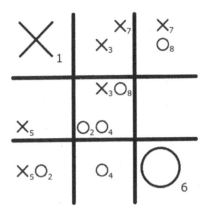

A game of Quantum Tic-Tac-Toe in progress. The second player has just made move O8. The first player must now choose whether to collapse)8 into the upper right square or the middle square. (Either way, O is going to get a tic-tac-toe.)

If you decide that the X1 over here is real, you erase the other X1. If there is only one mark left in that space, it just goes to that player, otherwise another decision may need to be made. In quantum terms, you are 'collapsing the wave function' and changing the probabilities to actualities. It has almost a Sudoku feel to it. Once a space gets a 'real' X or O, no more marks can be placed there.

Once you get the hang of it, it's pretty straightforward to play.

Back in Chapter 6, I talked about how many boys and girls you might guess there are in families with two children, and how things in physics could be correlated or 'entangled'.

One of the basic underpinnings of quantum physics is that things (whether particles, boys or girls, or cats) do not have specific properties until they are observed. They merely have probabilities of being in certain states, and then, when you look at them, they decide what they actually are.

The famous metaphor for this is 'Schrödinger's Cat'. Erwin Schrödinger was an early quantum theorist and proposed a thought experiment where a cat was trapped in a box with a radioactive source and a vial of poison gas. (Apparently, Schrödinger was not a fan of cats.) If the radioactive source emitted a particle, the vial broke and the cat died. You can't see in the box, but you wait an amount of time where you know there's a 50 per cent chance that the cat is still alive. Is it alive or dead? According to quantum mechanics, it is both. It is in a superposition of both of those states.

Einstein was not happy about this. It led to his famous quote, 'God does not throw dice.' He was convinced that the cat was definitely either alive or dead – it couldn't be in both states at the same time. So, he developed an alternative thought experiment to show that correlation and superposition couldn't be true. This became a famous paradox in physics called the Einstein–Podolsky–Rosen paradox. It goes as follows.

There are many physical processes where two particles are created at the same time and have related properties. For example, an electron and positron are created and fly apart from each other. One has negative charge, and one has positive charge – but you don't know which. Each has a fifty-

fifty chance of being plus or minus, and they are said to be 'correlated' or 'entangled'. Once you measure one of them, you know what the other has to be. This has been proven in many experiments.

Quantum mechanics says that when the two particles are created, each one isn't positive or negative; it's both. In the most common interpretation of quantum theory, they exist in a weird combination of both states until you look at them. And once you look at one of the particles, it randomly picks if it is positive or negative, and the other one instantly becomes the opposite.

So, Einstein proposed this thought experiment. Say you have two experimenters called Tom and Mandi, who are far apart from each other. The particles fly apart, one toward Tom and one toward Mandi. Tom measures his particle, and it is positive – so the partner particle must be negative. But if Mandi measures her particle, she will also see that it is the opposite charge of Tom's – no matter how close in time to Tom's measurement she makes it. This means that Tom's particle has to send the information about what to be to the other particle faster than the speed of light, which seems to violate the laws of relativity. Einstein said that the particles must 'know' what they are as soon as they separate. There is some 'hidden information' that we don't have access to, so they don't have to communicate across long distances instantaneously. So, quantum mechanics is wrong.

This paradox was proposed in 1935, the same year as Schrödinger's Cat, but no one could figure out how to tell if the particles really didn't know what they were until measured, or if they actually had a definite state as soon as they were created. No one, that is, until 1964, when physicist John Bell came up with the idea of looking at groups of correlated particles – in essence, correlations of correlations. He developed a theorem around this idea and designed an experiment that would distinguish between particles that have 'hidden information' about what they are, and those that just make it up when they need to.

Unfortunately, the experiment was too complex to be performed at that time – it couldn't be done for eight years, until 1972. And sure enough, it showed that Einstein was wrong. The particles 'decide' what they are when they are measured. It has been repeated many times since then, with an experiment in 2008 putting the measurements 18 kilometres apart to make sure there was no way information could travel in between – but the overwhelming evidence is that quantum theory is right. The particles do not decide what they are until they are measured.

This is highly weird.

Physicist Niels Bohr famously said, 'Those who are not shocked when they first come across quantum theory cannot possibly have understood it.' However, it is the most accurate theory humans have ever developed, and this concept of entanglement is being used to develop computers that have the potential to be tremendously more powerful than what we have now.

It would be difficult to fully explain Bell's Theorem in the space of this book. It is counterintuitive, and it's tough to create a metaphor or analogy that captures the idea. That's one reason why even Einstein didn't believe lots of the early results of quantum theory. But if you're interested in learning more, there are some online resources that get into the weeds.[2]

Okay, let's bring this back to the world of games. Even leaving quantum theory behind, entanglement and correlation can really mess with your intuitive sense of probability.

Let's say four people are playing *Bridge*. One of them says, 'I have an Ace,' and we know she is telling the truth. The chance that she's holding more than one Ace is about 37 per cent.

Later, the same player says, again truthfully, 'I have the Ace of Spades.' Strangely, the chance that she has more than one Ace is now 56 per cent.

When I first heard this, I didn't believe it. It makes no sense. But it's true. To understand it, I had to sit down with a small deck with just two Aces and two Kings. When you say, 'I have an Ace of Spades', you eliminate more hands than you do with just 'I have an Ace.' And a higher percentage of those hands contain a second Ace.

Be very careful with correlations – our intuition fails more often than not.

Chapter 14

When Math Doesn't Have All the Answers

This sentence is false.

Oh yeah? Prove it.

A set of game rules is a logical construction. The rules tell you what you are permitted to do, what you are prohibited from doing, and how the state of the game — scores, locations of pieces, etc. — all change as a result of the players taking an action. So, you can't have a rule that says playing a certain card gives you one point, and another rule that says that it gives you two. And you can't have a rule that says you are allowed to do something without explaining what the results of doing that will be. In logic, these two criteria are called consistency — that the rules do not contradict each other — and completeness — that all possibilities are discussed.

Mathematical Axioms

Mathematicians have tried to create logical constructions that encompass all mathematical truths. This type of logical system has two parts: axioms and rules for manipulating axioms.

An axiom is a statement that cannot be proved or disproved – it just has to be accepted. The Greek mathematician Euclid was the first one to develop the idea of an axiom when he developed geometry. For example, the idea that a line is the shortest distance between two points is an axiom of Euclidean geometry. It's the definition of a line. Another axiom says that a circle is defined by a centre point and a radius.

In addition to axioms, there are also rules about manipulating axioms. For example, if two things are equal to the same thing, they are equal to each other. If two things are equal to each other and the same thing is added to both of them, they are still equal. From these basic rules, you can derive new rules – like that the angles in a triangle all add up to 180 degrees.

This is very similar to board-game rules. As a simple example, in *Chess*, there is no explicit rule that says that bishops stay on the same colour spaces. But it arises naturally out of the basic rules. This is called 'emergent behaviour', and it features prominently not just in games, but also in computer science and biology, as simple rules can yield complex results.

In 1910, the British mathematicians Bertrand Russell and Alfred Whitehead published the three-volume *Principia*

Mathematica.[1] This was an attempt to place the entire foundation of mathematics on an axiom system, including traditional math, set theory and complex analysis. The idea was to transform mathematics into symbol manipulation. You would start with a series of symbols which represented one of the axioms. There were set rules that said how you could transform that string of symbols into a new string. For example, you could combine these two symbols and get another, or you could always add a certain symbol in a certain place.

These new strings of symbols would represent 'theorems' in the *Principia* system. There was a machine-like series of rules that could be followed to churn out new mathematical theorems, one after the other. Given enough time, a computer applying these rules one after the other would eventually be able to prove any theorem. The axioms and the rules that manipulated them were like seeds that contained within them all the truth and beauty inherent in mathematics.

Well, that was the idea anyway. In 1931, the mathematician Kurt Gödel published his now famous Incompleteness Theorem.[2] He proved that if a system of axioms and rules is complex enough that you *think* you can prove everything, there are true statements that you will never be able to prove, and falsehoods that you will never be able to disprove. Your system is fundamentally incomplete.

The proof is based on taking the statement 'This sentence is false' and translating it into the language of number theory. It is impossible to prove or disprove.

As soon as your set of rules becomes sophisticated enough to be able to express complex ideas, you have sown the seeds of your own destruction.

The ideas behind Gödel's Incompleteness Theorem have been extended to philosophise about the limits of computers, human intelligence, and even the idea of science itself, since all of these rely on rules to encode and process information. Are there truths that are beyond our capacity to grasp? Mathematicians get irate at these comparisons, because it is dangerous to extend Gödel's results beyond the rigid realm of logic ... but, still, it is interesting to ponder.

And so back to game rules. As game rules themselves get more complex, might they not also pass a threshold of complexity beyond which it is inevitable that situations will arise that are not covered? Is it impossible to make a very complex set of rules both consistent and complete? It certainly seems that way when you sit down to try to play a game of *Advanced Squad Leader* or *A World at War*, war simulation games that have hundred of pages of rules.

I think in those cases, however, the lack of consistency is due to combinatorial explosion rather than a Gödelian wall where completeness is inherently impossible. It is the same situation that leads to 'codex creep' or 'power creep' discussed back in Chapter 8, and similar in nature to the autocatalytic sets discussed in Chapter 10. For example, *Magic: The Gathering* has almost 12,000 different cards in existence. There are over a hundred million different pairs of two cards, and it is impossible to anticipate all the interactions. It is a statistical certainty that, in all those pairs, there will be some that lead to a runaway victory and need to be banned so that tournament play can remain competitive.

'NP Complete' Problems

> Given a map with several cities on it, find the shortest route that visits each city once.

This is known as the 'Travelling Salesman' problem, and while it seems intuitively simple, in fact, when you get above a handful of cities, the number of possible trips becomes huge. For example, if you look at visiting just 15 cities, there are over 40 billion possible trips.

Here's another one:

> Given a flat map, no matter how simple, how many colours are needed to ensure no two regions of the same colour ever touch (even diagonally)?

This is known as the 'Four-colour Map' problem, and its theorem proposes that, no matter how convoluted the map is, it needs at most four colours. The idea was first proposed in 1852. It was quickly proved that five colours would always do the job, and there were examples that showed that three colours would not be enough — but four colours were in a limbo. There was no proof that four would always work, but also no counter-example showing that they wouldn't.

Because it was so easy to state, the Four-colour Map

Theorem captured the imaginations of mathematicians for over 100 years as they laboured to solve it. It was finally conquered in 1972 and has the honour of being the first major proof solved by a computer. Mathematicians had the computer analyse thousands of key maps to prove the conjecture.

So we know that any map can be coloured with four different colours. But another, harder form of this problem remains. If I give you an arbitrary map, figure out if it can be coloured with three colours or not.

Both the Travelling Salesman and Three-colour Map problems are a special type of problem called NP Complete. 'NP' stands for 'non-polynomial'. This means that the length of time required to find the best answer to the problem – like the shortest Travelling Salesman trip – grows really big, really quickly as the number of items grows – like the 40 billion possible solutions for the 15-city trip. The solution set gets large to the point where even moderately sized NP Complete problems would take billions of years to solve on today's computers.

Another feature of NP Complete problems is that it is easy to tell if a solution meets the requirements of the problem, even if it is not optimal. For example, it is very easy to tell if a proposed answer to the Travelling Salesman problem really visits each city one time. So, NP Complete problems are easy to state and easy to find answers for, but devilishly hard to find the *best* answer for.

There are a lot of games that are based on NP Complete problems, including several that are based on the Travelling

Salesman problem. The excellent *Elfenland* by designer Alan Moon is perhaps the most famous. But any pick-up and deliver game, like *Thurn and Taxis* or *Empire Builder*, have elements of this NP Complete problem, as players have to optimise their routes. Echoes of the Four-colour Map problem are found in *Blokus*, *Take It Easy*, and other tile-placing games. You are trying to place tiles to satisfy certain constraints in an optimum way.

Another famous problem is the 'Knapsack' problem. Like the others, it is easy to describe. You are given a bunch of objects of different size and values, and a knapsack that can only hold objects of a certain total size. The problem is to find the combination of objects that fit in the knapsack and have the highest total value. The game *Pack & Stack* by publisher Mayfair, which has you trying to load different value blocks into your ute, is based on this problem. I think the problem could also be used as a cool mechanic in future games.

In addition to games, NP Complete problems are fundamental to a variety of practical tasks. Logistics companies like FedEx spend a lot of time working on the Travelling Salesman, and the Knapsack problem is the basis of a lot of the encryption algorithms that are used to protect data. So, any optimisation to finding solutions to these problems will be huge. So far, some minor optimisations have been developed by computer scientists and mathematicians, but they are still really difficult at their core, and work on solving NP Complete problems continues.

But there is another impetus for working on these problems. The problems have an intriguing property: they are all the same! If you figure out how to solve one, you can solve them all. How is this possible?

Two seemingly unrelated NP Complete problems can be mapped onto each other by various math tricks. Let me illustrate through – not surprisingly – a game.

Here's a two-player game called *15*. Take your deck of cards and lay out your Clubs from Ace (One) to Nine. In turn, each player takes a card. Keep going until one player wins by collecting a set of three Clubs that total exactly 15 or until all numbers are gone, which is a draw. For example, Four, Nine and Two will win. You can try it if you'd like – see if you can figure out a strategy.

But here's another way to look at this game. Take those nine cards and lay them out in a 3 x 3 grid, where each row, column and diagonal adds up to 15. This is called a 'magic square' and the Five of Clubs always has to go in the middle (see overleaf).

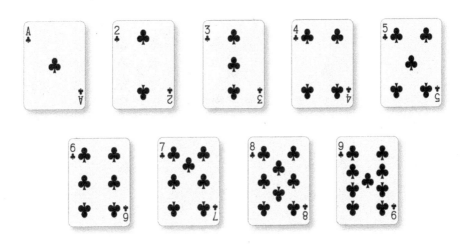

So, we have just transformed this problem. With a magic square, picking three cards equalling 15 is the exact same game as *Tic-Tac-Toe*. Now it's easy to figure out what your strategy is — you have transformed *15* into something you know how to win.

Same thing with NP Complete problems. Through clever mappings, you can change any of them into any other. So, if you can figure out a clever way to solve the Travelling Salesman problem, you know how to solve the Knapsack problem, and any of the known NP Complete problems.

And, in addition to gaining fame and fortune, you'll be good at a bunch of games.

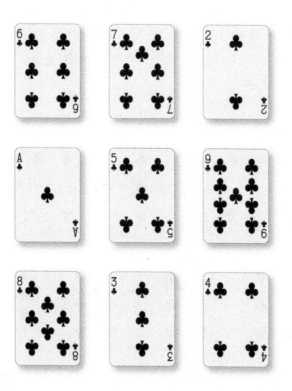

A Butterfly Flaps its Wings

In 1963, mathematician Edward Lorenz published a paper examining what would come to be known as 'chaos theory' and the 'butterfly effect'. Contrary to popular culture, Lorenz's paper 'Does the Flap of a Butterfly's Wings in Brazil Set Off a Tornado in Texas?' is not a biological treatise on the tornado-causing powers of South American insects. Rather, it looks at how tiny changes in the input of a system (like the flap of a butterfly's wings) lead to wildly different outputs (in this case, a tornado), even if the system itself is completely deterministic, with no randomness whatsoever.

To understand this, however, we need to go back another 300 years to 1687, when Isaac Newton had just published *Philosophiæ Naturalis Principia Mathematica*. Among other things, this laid the foundations for Newton's theory of gravity. He showed that if you have two objects acting under gravity, they will orbit each other in ellipses or, if they are going really fast, hyperbolas. Newton's equations matched the data beautifully, and laid the foundation for the idea of the 'clockwork universe' – which suggested that if you know the initial conditions of the universe, you can calculate exactly what will happen for every moment in time going forwards. This idea perhaps reached its pinnacle in 1846, when astronomer Alexis Bouvard noticed that there was something weird about the orbit of Uranus – something that would only make sense if there were another planet out there in a particular location. Bouvard went out looking for it and, sure enough, there was the hitherto undiscovered Neptune.

However, there were still a few loose ends. One of those is what is called the 'Three-body' problem. If there are two objects moving under gravity, you can write down an exact equation for their motion. But if there are three objects – say the Earth, Sun and Moon? Then, not so much. It actually turns out that it is impossible to devise an equation that describes their motions. You have to use a series of equations that approximate it. And if you can't even do three bodies, what hope is there for describing the whole universe?

'Well,' the physicist would answer, 'we may not be able to get an exact equation, but we can use math to get really, really close to the answer – so we, in principle, could figure it out.'

That turned out not to be correct either. In 1880, Henri Poincaré was studying the orbital patterns of three bodies and noticed that if you change the starting positions just a teeny tiny bit, it could make a really big difference in the ending positions. This is not because of randomness – we are not talking about quantum mechanics where randomness is part of it. It is, as Lorenz said, due to tiny changes in the input of the system, which are impossible to measure.

This is where chaos theory comes from. Because there will always be some slight error in your initial positions, the evolution of a system is very unpredictable.

So, what does chaos theory have to do with games? There are definitely games that exhibit this behaviour – where small changes in starting positions, or small moves, can make big differences in the results. 'Abstract' strategy games certainly have this feature. In the game *Go*, players place stones on the board to surround territory and capture opposing stones. When the position becomes complex, placing a stone on a certain intersection, or on an intersection that is right next to it, can have a very unpredictable result on the game — even though theoretically you can work through the exact implications.

Another example is the game *Girl Genius* by designer James Earnest. In this game, you lay out a grid of cards face down on the table. On your turn, you flip up a new card and then rotate another face-up card 180 degrees and follow the instructions on it, which usually results in some effect on neighbouring cards. These effects may trigger other effects and so on, as they cascade around the grid.

Other games have this feature as well, like *Robo Rally*, where you program the movement of your robot in predictable ways, but the different board elements (such as other players' movements pushing your robot out of place) can put your robot far from where you thought it would end up. Small changes to the inputs, big changes to the outputs.

I personally like chaos as a design element — and, again, I mean deterministic chaos, not random chaos. I think it adds an element of fun, and can also lead to those moments of clarity, where all of a sudden the path to victory crystallises in front of you in the middle of the chaos of possibilities.

But you don't see it as a design element that often. One reason is that it can certainly lead to analysis paralysis, because the results of all of these interactions can theoretically be determined by a player if they spend enough time. It can also make it difficult, or impossible, to have a long-term strategy, since the game situation can change so radically from one turn to the next. Also, it is challenging to handle as a designer, since figuring out if the game is really balanced or not is that much more difficult when it can go careening in different directions.

But I guess that all games exhibit a sort of 'slow-motion' chaos over the course of the entire game, as player choices, all of which may be logical and not that different from the last time you played, push the game into radically different directions. And choices you made early on can come back to haunt you in unexpected ways.

So, as you're playing, keep an eye out for chaotic elements and think about how they impact the design of the game and your play experience.

Graph Theory and Train Games

There are a lot of different types of train games — pick up and deliver, stock manipulation, technology improvement — but almost all involve connecting different places on a map.

This has a direct allegory in the area of mathematics called 'graph theory', which is the study of networks (series of nodes/dots connected by paths/lines), and is of fundamental importance in NP Complete problems like the Travelling Salesman.

Graph theory got its start in 1736 in a paper written by one of the great mathematicians of all time, Leonhard Euler of Switzerland. Euler did it by tackling the famous 'Seven Bridges of Königsberg' problem.[2]

Königsberg was a city in Prussia on the Pregel River.

There were two large islands in the middle of the city that were connected to the city and each other by seven bridges. The problem was to try to find a path that went over every bridge, but each bridge being crossed only one time.

Euler was able to demonstrate that it was impossible to do. He proved that the problem was equivalent to a map of nodes and lines, and that you could only trace a complete path through such a graph if there were no more than two nodes with an odd number of connections. This is equivalent to the children's puzzles where you have to trace a figure without lifting your pencil off the paper, and without going over the same line twice.

Graph theory has gone in a thousand different directions over nearly three centuries, and is used to study the internet, social networks, and disease propagation, among other topics.

But here I will be discussing … hold onto your hats … Erdös–Rényi random graphs. And how they relate to train games (especially games like *TransAmerica*) and life in general.

Here's how you make an Erdös–Rényi random graph. Take a blank piece of paper and put a bunch of dots on it. Now randomly pick two dots and connect them with a line. Then do it again. Keep doing this over and over again, until all of the dots are connected into a single blob.

Now here's the interesting part. Each time you connect a pair of dots, find the largest group of dots that are connected and record how many dots are in that group.

What you'll find is that the size of the largest group starts out small and stays small for quite some time. In fact, if the number of connections

When there are only a few connections, the groups are disconnected. The largest only have one or two nodes.

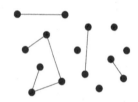

When the number of lines are right at the critical point, larger groups suddenly appear.

As the number of connections continue to increase, the largest group contains almost all the nodes

is slightly less than half the number of dots, the largest group will almost certainly be pretty small. But once the number of connections is slightly larger than half the number of dots, there arises a single, dominant, large group of dots that will eventually absorb all of the dots in the graph.

So, the curve showing the largest group of dots has an S shape, with the steep rise right when you reach the halfway point.

In science, this is called a 'phase transition' – a sudden change of a system from one state to another. Just as water changes from a liquid to ice quite suddenly, so too does the graph change from a liquid state (where there are lots of small groups) to a solid state (where there is one giant clump of dots).

It's amazing (and a little cool) that a simple system like this – dots and lines – can exhibit such interesting and unexpected behaviour.

You will see similar behaviour in many train games, where cities are connected by paths. In *TransAmerica*, for example, the players place tracks onto a map of the United States, trying to connect their five secret destination cities into one network. But all the players share the same track, so you want to try to take advantage of what your opponents do.

TransAmerica is, admittedly, a very light game. But it does have this interesting property of a phase transition. At first, players work on independent segments of tracks. But then, eventually, you have to make the decision of when to merge the tracks together. And it seems to happen all at the same time, as players make a last effort to reach their targets. There is indeed a phase transition in the nature of the connecting tracks about halfway through the game. The tracks change from a 'liquid' separate state, to a solid clump of tracks.

So, the next time you are playing a connection game, whether it's a train game, or some other theme like the hotels of *Acquire*, think about graph theory, and the phase transition between small groups and a super-connected network. It will give you a new appreciation for the phases of the game, and new insight into how to predict what's going to happen.

Plus, you'll get to say to your friends, 'This is just like Erdös–Rényi random graph theory!'

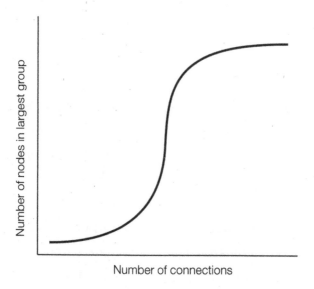

As the number of connections grows, there is a big jump in the size of the largest group when the number of connections is around half the number of nodes.

Chapter 15

Goldilocks, Three Bears and a Whole Lot of Noise

Too Much Choice or Just Right?

If you recall from Chapter 4, gamers are most interested in games that offer meaningful choices, maybe with a bit of luck thrown in. We want to be masters of our own destiny, but as the Fox and the Cat demonstrated, too much choice can sometimes be overwhelming.

The more choices you offer players in your game, the more difficult it becomes for players to make a choice, the lower their quality of choice, and the more prone to analysis paralysis they will be. There is evidence from *Chess* that the more possibilities Grandmasters are considering, the more likely it is that they will commit a blunder.[1] However, the fewer choices you offer players in your game, the more your game feels unfairly predetermined, or comes down to a literal roll of the dice.

Just like Goldilocks rampaging through the home of the three bears, you want the level of choice in your game to be 'just right'.

*

Of course, that's not all that is going on when you ask a player to make a choice.

Let me explain. I'll describe two games, and you pick which of the two you would prefer to play.

In both games, you have to guess if a die roll is even or odd. If you're right, you win $20. If you're wrong, you just walk away.

For the first game, you make your guess, then roll the die. For the second game, I roll the die in a dice cup and slam it down onto the table, and then you make your guess.

In game A, you guess and then roll. In game B, you guess whether a die that has already been rolled is even or odd.

Which would you rather play?

Statistically, of course, there is absolutely no difference between these two games. You have a fifty-fifty chance of winning or losing.

However, most people prefer to play Game A, and bet on a future roll. In fact, two-thirds of people who are asked this question choose to guess something that is happening in the future, not the past.[2]

Here's another example — an experiment that was performed in 1961 by economist Daniel Ellsberg.[3]

Subjects are presented with two boxes. One has 50 red balls and 50 green balls. They can open the box and count if they like. The second box has an unknown mix of red and green balls, but has at least one ball in it.

The subject's job is to select a box and name a colour, either red or green. If a random ball pulled out of the box matches their chosen colour, they win $20. If they lose, they walk away with nothing.

So, which would you pick? The fifty-fifty box, or the unknown box? In this experiment, an overwhelming majority of players selected the box with a known fifty-fifty split. I'm guessing you did as well.

Now, mathematically, there is absolutely no difference between the two boxes, because you can name your own colour. There is no way the experimenter can trick you or stack the deck in their favour. Your expected winnings are exactly the same whether you pick the known fifty-fifty box or the unknown box. Think about it for a few minutes if you're not convinced.

What's going on here? Is there a connection between these two experiments?

There were a variety of theories proposed, but they all had deficiencies – that was until 1991, when psychologists Amos Tversky (who, you may recall, developed prospect theory with Daniel Kahneman) and Chip Heath proposed a framework they called 'competence'.[4]

The theory says that people prefer to make bets or play games where they have a higher level of 'competence' – meaning, for this purpose, the amount that you know about something as a percentage of what can ever be known.

Let's go back to our first example, the die roll in the past or the future. If the die roll is in the past, there's a key piece of information that is potentially available (what was rolled) that you don't know. So, your competence is low. In the future case, you can't possibly know what the die roll is going to be, and neither can anyone else, so your competence is higher.

In the second game, you know everything there is to know about the fifty-fifty box. You know exactly what the chances are of each colour. But in the unknown box, even though your expectation value is exactly the same (to win 50 per cent of the time), your competence is very low. The experimenter has a key fact about the world (the red/green split in the box) that you don't have.

In both cases, you will tend to go for the choice where you have a higher competence.

There are several theories on why we like to go with the higher competence choice. I like the 'blame' theory. Competence helps people take credit when they win, as it seems like they knew what they were doing, and it helps avoid blame when they lose – players can always point to all the reasons why their choice was 'smart', even if it turned out to be wrong. If players are operating with less information (lower competence) and they lose, their choice can be made out to be foolish, whereas if they win, it can be attributed simply to dumb luck.

A game designer can use this to skew decision-making in subtle and interesting ways. For example, let's say we have two players, Susan and Eric, playing a fighting game where the cards are first placed face down before being flipped after both players have made a choice. Susan is playing an attack card, and Eric has to try to play a block card that hopefully matches Susan's. You could design it so that Susan has to place an attack card first, and then Eric places the block card. Or you could make it so Eric places the block card first, and then Susan places her attack card afterward.

I think you'd find that the player who has to place the second card, responding to what their opponent played secretly, is going to be much more prone to uncertainty and extended analysis, even though the chances of matching are the same either way.

Too Much Luck or Just Right?

Meaningful choice may be well and good in a game, but Goldilocks also wants a little bit of uncertainty in her games. Too much luck, and a game can feel unpredictable, defeating the purpose of choice. Too little luck, and the game can get bogged down in excessive analysis paralysis, as players struggle over the multitude of choices available to them (unless, of course, you are designing a wholly abstract game like *Chess* or *Santorini*, in which luck plays no part).

You want your game to have just enough luck. But how much luck is just right?

To answer that question, I am going to look at randomness through the lens of something those in the science world call 'noise'. You might have heard of it.

Back in the day, if you tuned a TV to a channel that wasn't broadcasting, you would see a screen filled with a crackling black-and-white pattern. Now, of course, it just says 'No Signal' or something equally boring. But back then, these fluctuating patterns on the TV were weirdly mesmerising – they were even used rather memorably in the 1982 film *Poltergeist* as a way to communicate with the spirits of the dead.

This pattern is noise that is caused on the screen by fluctuations in the local radiation field, including cosmic radiation from space. But it is not just noise; it is a special type of noise called 'white noise', and it has been used effectively in places such as randomisers in computer chips, as we saw back in Chapter 9.

White noise is a random pattern where each pixel on the screen is not influenced at all by adjacent pixels. The technical term for this influence is 'correlation'. There is no correlation between any dots on the display. Each one can display any value at all.

White noise, as its name implies, can also be expressed as a sound. The lack of correlation means that the frequency at one point is completely unconnected with the frequency at the next point. It just jumps around randomly.

If you are composing music that is like white noise (sure to make you rich and famous), one way to do it is to roll, say, a twenty-sided die, and have each number correspond to a note. You just roll the die over and over, and record the notes in sequence. This sequence is completely random and will just jump around. It will not be particularly fun to listen to.

White noise is not the only type of noise you can have. You can have noise where there actually is some correlation between one moment and the next. One type is called 'brown noise'. Here's a way you can produce it with a six-sided die. Start with the number ten. Now roll the six-sided die. If it's a 1 or a 2, go down by one note. If it's a 3 or 4, stay the same. If it's a 5 or a 6, increase by one. So if you roll a 1, your ten becomes a nine. If you then roll a 5 you go back up to ten.

This is still a random process, but now it is highly correlated. The last value has a very strong impact on what the next value can be. You can't just jump from a five to a 12 with brown noise – you need to pass through steps in between. This process is sometimes called a 'random walk', and occurs in nature as 'Brownian motion' – which is where the name 'brown noise' comes from.

'Brownian Motion' is named for botanist Robert Brown, who first wrote about it in 1827. He was examining pollen grains suspended in water, and under a microscope observed that tiny particles that broke off of the main grain would jitter back and forth, seemingly at random. At first, he thought it might be a sign of life. So he tested with inorganic material and when he saw the same motion, he realised it must be a general property of particles in a liquid. Scientists assumed, correctly, that it was related to atoms and molecules randomly bouncing off of the tiny particles. But a full mathematical description wasn't presented until Albert Einstein did so in 1905.

If you were to have brown noise on your TV, unlike the jumpy static pattern of white noise, you'd have slow, undulating areas of colour.

So, white noise is completely uncorrelated and brown noise is extremely correlated. There must be a middle ground, right? Well, there is – and this is where it gets fun. There is a type of noise that is partially correlated. To continue our music example, with this noise it is most likely that the next note will be close to the prior note, but there is also a chance that it will jump far away.

This type of noise is called '1/f noise', or 'pink noise'. In pink noise, the size of a jump is related to how often it happens. Small jumps happen a lot – and the bigger the jump, the less likely it is to occur.

It turns out that if you take normal music (music that we enjoy listening to, unlike randomly generated white or brown music) and you plot the intervals between notes and their frequency, just about every song is close to true pink noise. Same thing with art. Lots of small jumps and fewer big jumps.

The pink noise has more texture to it. It's not musical, obviously, but it is more interesting to listen to than white or brown noise.

Can you generate pink noise with dice? Yes, actually. Here's one method. Let's say you have four dice: A, B, C and D. At the start of the game, you roll all four dice and add them up – that's your starting point. Now, every turn, you re-roll die A and recalculate your sum. Every other turn, you also roll die B. Every fourth turn, you also roll die C, and every eight turns, you add in die D and roll all four dice. Most of the time your changes will be small, but periodically you have the chance for a larger change.

So, what does this have to do with games? Well, many games have random processes. These can be white, brown or pink, and they all have different effects on gameplay.

One way to think about noise in games is by asking how easy it is to predict what the next value is going to be. In a game that uses white noise, there is no way to predict the result. If the last number was a 12, the next number could be a five or a 72 with equal probability. There is no 'memory' in the system.

Schoko & Co is a game about managing a chocolate factory. One of the key factors is how much chocolate the market wants. This is randomly determined by drawing from a deck of cards, which tells you exactly how many units are desired, and the minimum and maximum price.

This is white noise. There is absolutely no correlation from one round to the next. And *Schoko* suffers as a game because of this. You are at the whim of the market, and if you guess that there will be big demand but there isn't … tough luck. White noise randomness very much takes any sense of meaningful choice away from the players.

Brown noise is the exact opposite. It's fairly easy to predict what the next value is going to be. If the last number was 12, the next number may be 11 or 13, but certainly won't be 72.

In *Evo*, you control a group of dinosaurs trying to survive as the climate changes from hot to cold and back again along a climate track. Each turn, you roll a die. Mostly, the temperature becomes colder as the game goes on, but it might stay in the same spot or even get hotter again.

This is brown noise. The possible next spaces are very tightly controlled and you can make reasonable predictions about what will be happening next turn. It makes *Evo* seem a little flat, and I always find myself a bit frustrated by the climate track. Given the length of the game, there just isn't that much movement on it.

Another game that had a similar issue was the original 1974 version of *Crude*, published by St Laurent Games. A key aspect of the game was the current type of economy. It could be in one of seven different states, like Depression, Recession or Expansion. The economy changed when you rolled doubles, then you rolled a die to see what the next economic state was. Most likely it would move along a standard path, but you might move back to an earlier step. This is very similar to *Evo* in that the movement of the economy was brown noise, but because it didn't happen every turn it had a bit more chaos to it. Unfortunately, this wasn't good chaos: since the economy changed when you rolled doubles, the amount of time between changes was white noise. So, it had a similar too-out-of-control feel.

Pink noise is in the middle. Most of the values will be close to the prior value, but some may jump away. The further a jump is, the less likely it is to happen. As mentioned, people find sequences like this the most interesting. Intervals in music follow this pattern, and it applies across a wide variety of disciplines.

In Stronghold Games's 2012 reprint of *Crude*, the economic change system is modified. Now it doesn't change on doubles. You roll two dice, and the difference between the dice (that is, a number from 0 to 5) is added to a marker on a track. When the track reaches 8, the economy changes. The movement of this marker is a pink noise process. It is more likely to move by small amounts, but there is a chance it may move a lot if you roll a 1 and a 6. So, you can see economic changes coming, but you don't know exactly how long they will take.

However, the actual next step that the economy moves to is still brown noise, so perhaps this is a missed opportunity to add more controlled chaos into the game.

The last game I'd like to discuss is the dinosaur game *Tyranno Ex* – another game with a changing environment. But here it changes in a very interesting way. The environment is represented by face-up discs of various types. Players can add face-down discs to the climate track, and under certain conditions they are flipped over and the environment may change. But there is an ingenious cascade mechanism that may result in several environmental changes at the same time. While you may have placed a lot of Sun discs in an environment, they may still get swept aside. So, players have some control over what's going on, but there is a chaotic element as well that can thwart your plans.

This is pink noise, and you can see the way that it makes a game much more interesting. It also makes use of another element of chaos theory, discussed back in Chapter 14. The 'cascade' mechanism offers a form of the butterfly effect, where small changes in the input of the system (in this case, placing a specific kind of disc onto the climate track) can result in large, unpredictable outcomes. This does exactly what I was talking about back in that chapter, providing randomness balanced with the opportunity for singular moments of clarity that can swing a game in a player's favour.

Goldilocks might sit down to a game of *Tyranno Ex* and consider it just right.

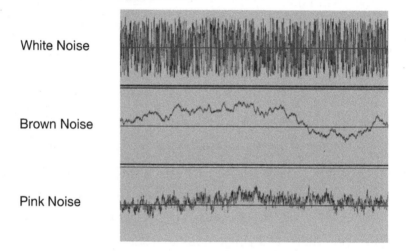

Chapter 16

Trading to Get What You Want

A trade is an exchange of value between parties.

That seems simple enough, but it also raises the questions 'What is value? How do we measure it?' The answer to that is at the heart of any trading game. The value traded can be *direct* – where you are trading in what ultimately determines the victor of the game – or *indirect* – where you are trading in resources that assist you in gaining whatever it is you need to win.

Pit, for example, is a *direct* trading game. Your goal is to collect all of a particular type of card. Here you are dealing directly in the currency of victory. If you're collecting Oranges and you get three Orange cards in a trade, you are three points closer to victory.

In an *indirect* trading game, on the other hand, like *Settlers of Catan*, you are trading resources. These can then be used to build towns or cities, which do yield victory points, or build roads or buy cards, which may get victory points. You can also use them as sources for future trades. So, the trading is one step removed from the actual victory.

An even more indirect trading game is *Bohnanza*. In this game, you can actually trade something for nothing: just give a card to someone. So, a trade doesn't even have to involve property. It might be something intangible, like freeing your hand of a bad card, or just trading for good will and/or to establish a good reputation.

When you are trading, you want to get back at least as much value as you are giving up. If I am giving up something that can give me three points of value, I want to get at least three points of value back. There is no point in making the trade if that is not the case.

So, if I am giving another player something that is worth three points of value and she gives me back something worth three points of value, why should we make the trade at all?

The short answer is — you shouldn't. There is no reason. The long answer: a key design feature of trading games is that they are asymmetric — you may be giving up three points of value, but that same thing may be worth five points to the person who is receiving it. Without asymmetry, it is very difficult to make a compelling trading game.

And the basic weapon in the game designer's arsenal to create asymmetry is the concept of 'set collection' in a game. If you can group things in a certain way, they are worth more than they are individually. The whole is worth more than the sum of its parts.

In *Monopoly*, without the concept of the colour-group monopoly, there would be no reason to trade. In *Settlers of Catan*, if you weren't trying to put together particular sets to buy things, there would be no reason to trade. In *Civilization*, if groups of commodities did not rapidly escalate in value based on how many you have, there would be no reason to trade. Introducing asymmetry creates the energy behind the trading engine.

Pit is the closest game that I can think of that has symmetric trading. To overcome this symmetry, the designer added the real-time feature – and the screaming and the time pressure to add energy. Turn-based, slow, deliberate *Pit* would be no fun at all.

Asymmetry is key in most trading games, but, while it is almost always necessary, it is not sufficient.

Let's say I'm giving you something worth three to me but six to you, and in exchange I'm getting something that is worth three to you but only five to me. Would I make this trade?

Well, in the real world I would, and we do all the time when we exchange goods and services. We would both be better off. And maybe if it happened over and over again, I'd start to get annoyed and take up the pitchforks and storm the castle – but we're talking about games.

In a game, there are winners and losers, and even though we both move up, this is a better deal for you. So, most of the time, I wouldn't take it.

Now, there are circumstances where you might want to. If you're ahead of the other player, you might take it. If a third player is way ahead, you might take it so you both get a little closer.

But if things are even, it's not a great deal for you, and makes it more likely your opponent will win. You need to look not just at the value you gain, but also the value your opponent gains.

So most great trading games also have hidden information. This can be done in many ways. Some games give you a hand of hidden cards, like *Bohnanza*. It is easy to tell what a trade will do for your position, but hard to tell how much it will help your opponent.

In other games, information is hidden in the future. In *Monopoly*, for example, it can make a big difference to the ultimate value of your holdings if your opponent lands on your monopoly before you land on theirs. In *Settlers of Catan*, a severe rock shortage can turn into a rock surplus in a few turns, making you feel silly for giving up three resources for that one rock.

On the other hand, you don't want the value to either you or your opponent to be so hidden and obscure that you can't make a judgment on it. That would take any skill or player interest right out of the design.

Revisiting Expectation Value

All of this comes back to how easy it is for a player to calculate the 'expectation value' of their trade. In particular, in *indirect* trading games, you must be able to determine how valuable a resource (or part of a set) is worth to you and to your opponent. Doing so allows you to play with all the information. In Chapter 5 I discussed the game *Masters of Commerce*. In that game, you were making deals with other players for properties, and then a roll of the die would determine their final value.

But by calculating the expectation value of the properties I was trading, I was better able to value my trades and come in first by a long way.

This is a classic example of too much visibility in expectation value and trading. If all the players in *Masters of Commerce* had used my strategy, there wouldn't have been much action. There would have been a 'known' value for buyers and sellers. A player could try to deviate from that, and they might get lucky and win that game, but in the long run they would lose more often.

Game designers need to have a way, especially in an auction or trading game, to obscure the expectation value of whatever is up for grabs. One way is to make items worth different amounts to different players. Maybe Bob really needs that item to make a valuable set. How much is it worth for you to block him? How much can you drive up the price before he sticks you with it? Or maybe there are so many random elements down the line that the expectation value is murky at best.

An example of something that could have pushed me away from the expectation value in *Masters of Commerce* might have been the introduction of inside information about price movement. Let's say that each player got one or two cards that moved the value of a particular property up or down one or two boxes in addition to the die being rolled. Now, Sarah may know that red properties are going to go up two boxes before the die is rolled, so she can factor that into her expectation value for red. But if I see that Sarah is willing to pay more than I would normally expect just based on the die roll, I may deduce that she has a card that will move red upward, and be willing to pay more myself. But she could just be bluffing, trying to sway the market and planning to get out of red just before the two-minute negotiations are up.

Just a small addition like this can add enough of a twist and psychological element to shake players away from the optimal strategy of expectation value. Elements of a game that force players to make judgments based on incomplete information lead to more interesting decisions and more involving games.

Triangular Numbers

What do the numbers 1, 3, 6, 10 and 15 have in common?

They are all what are called 'triangular numbers', and they can be useful when designing a game that relies on 'set collection' for scoring. Imagine a series of rows of dots. There is one dot in the first row. Then you add two dots in the second row, three dots in the third row, four dots in the fourth row, and so on. The dots will form a triangle, and match the way bowling pins are arranged. The number of dots grows by the sequence 1, 3, 6, 10, 15, 21 … and so on, as more rows are added.

In *Coloretto* you are collecting cards of different colours, and this particular sequence is used for final scoring. If you have three cards of the same colour, you get 6 points. A set of four cards nets you ten points.

In the game *Thebes*, you are an archaeologist trying to find artefacts to score victory points. But you can also get points by collecting 'Congress' cards as you travel through Europe.

At the end of the game, you score based on the triangular numbers: 1 point for one card, 3 points for two cards, 6 points for three cards, and so on.

This same mechanic is found in the winner of the first Spiel des Jahres (the famous Game of the Year award given out in Germany), the terrific *Hare and Tortoise*. In that game, you are racing other rabbits down a track. It looks like a 'roll-and-move' game like *Snakes and Ladders*: on your turn, you move forward and do an action based on the space you land on. But rather than rolling dice, you get to choose how many spaces forward you will move – the further you move, the more carrots you have to pay. And the cost in carrots is (you guessed it) a triangular number based on the number of spaces you want to move. If you want to go four spaces, you pay ten carrots. The reference card goes all the way up to moving 44 spaces, which takes an epic 990 carrots!

So why are triangular numbers so popular with game designers? When used in a positive way (for example, the scoring in *Coloretto*), they are an easy way to make the same resource worth different amounts to different players. This introduces complexity and tough decisions into what otherwise might be a simple mechanic. In *Thebes*, for example, a Congress card may only be worth one point to you, but if your opponent has six Congress cards, then it's worth seven points to her. So, you need to consider putting your own plans on hold to grab it just to deny her those points. If all of these cards were worth a fixed number of points, that decision would be much simpler to make.

There are a number of sequences that would create the same effect – perfect squares, or powers of two. But triangular numbers go up just fast enough to create this effect, without increasing so quickly that it dominates the game.

On the negative side, like paying more carrots to move more spaces in *Hare and Tortoise*, the designer is able to make you decide between distance and time. The most efficient move from a carrot standpoint is to move one space per turn and pay one carrot per turn. But you can't win by doing that. You need to balance carrots and time – move fast enough to stay ahead, but not so fast that you run out of carrots.

One last example – in *Ticket to Ride*, you score points when you claim train track segments as well as for connecting different cities together. The sequence of points for claiming segments from one to six spaces is 1, 2, 4, 7, 10, 15. This is suspiciously close to our triangular numbers sequence of 1, 3, 6, 10, 15, so I was wondering if the design originally used a more traditional triangular number sequence that may have changed through playtesting.

I will finish this discussion on triangular numbers and intransitive relationships with words from *Ticket to Ride*'s designer, Alan Moon:

> I like to tell people that I don't use mathematical formulas when I design games. But of course, that's not completely true. I know there are designers out there who start with a formula and create the game around it. I think that results in a dry game.

I start with mechanics/theme/bits/whatever, and then use formulas only if and when necessary. During playtesting, I'm trying to see how much fun the game is, and not if the math works. So, the scoring system of *Ticket to Ride* was made to fit the game, not the other way around. Any relation it has to any sequence is purely coincidental.

Intransitive Relationships

If I were to play a game of *Chess* against the reigning champion Magnus Carlsen, I would lose. I think even if he got struck by lightning halfway through the game, I would still lose.

And while I am not quite as certain, I am pretty sure that if I played *Chess* against my 11-year-old nephew, Isaac, I would win. At least for a few more years.

So, logically, it is reasonable for me to say that Grandmaster Carlsen would be able to beat my nephew.

If A can beat B, and B can beat C, then A can beat C. This type of relationship has a special name: it is called a 'transitive relationship', and it comes up all the time in different ways in mathematical proofs. For example, if B is a subset of A, and C is a subset of B, then C is also a subset of A.

Transitive relationships are all around us, and are constantly seen in game designs. Going back to *Chess*, a rook is stronger than a pawn, and a queen is stronger than a rook. So, a queen is stronger than a pawn. A ten-dollar note is ten times as valuable as a dollar coin, which is ten times as valuable as a ten-cent coin. A ten-dollar note is always more valuable than a ten-cent coin.

However, there are also game design elements that do not follow this rule – there is no 'best' element. There is always something that can beat it. This is called a non-transitive or intransitive relationship.

The most basic example is *Rock-Paper-Scissors*. Rock beats Scissors, and Scissors beats Paper, but Rock does not beat Paper. There is no dominant element.

In general, intransitive elements add a lot of interest to a game. For example, in the 1910 game *Stratego*, players control an army ranging from the mighty Marshal to the lowly Spy, all trying to capture their opponent's flag. The best piece is the Marshal, which can capture anything (it used to be a 1 when I was a kid, but now I hear it's a 10). And the lowest-ranking piece is the Spy. The Spy can't capture any piece on the board … except the Marshal. And adding this little twist, that little extra bit of intransitivity, really makes a huge difference in *Stratego*. If the Marshal were invulnerable to attack, the game would basically be pointless.

From a game-design perspective, having intransitive elements is an important part of making an interesting game by avoiding dominant strategies. Intransitivity can be added on a number of levels. The simplest is by making components that interact with each other in different ways, like *Stratego*. A lot of battle games take this approach. For example, the designer might make Infantry bad against Cavalry, Cavalry bad against Artillery, and Artillery bad against Infantry. So, it forces you to have a balanced force.

We used that approach in my game *The Ares Project*: every unit has some targets that it is good against and some that it is bad against; so, matching up is important.

A game can be intransitive on the strategic level as well. For example, in *Magic: The Gathering* there are three broad types of decks. Rush decks try to win the game as quickly as possible. Control decks play for the long game. And midrange decks are somewhere in between. In general, rush decks beat control decks, control decks beat midrange decks, and midrange decks beat rush decks. These intransitive strategies can arise out of transitive elements in the game. Individual *Magic* cards may be strictly better or worse than another card, but in an entire strategy there is no 'best' card.

There is one particularly cool way to add intransitive elements to a game: with dice. There is a way to make a set of dice – say three dice, A, B, and C – so that if you roll A and B, A will roll higher more of the time, and if you roll B against C, B will tend to roll higher, but if you roll C against A, then C will tend to roll higher. These are called nontransitive dice.

There is a set of four nontransitive dice called Efron Dice, after the inventor, Bradley Efron, that increases the winning chances to two-thirds if a die is rolled against its 'good' die.

Warren Buffett used to carry a set of these dice around with him and use them for a bar bet. Bill Gates has told a story that Buffett brought the dice out once when they were at a meeting. Buffett suggested that each of them choose one of the dice, then discard the other two. They would bet on who would roll the highest number most often. Buffett offered to let Gates pick his first. This suggestion instantly aroused Gates's curiosity. He asked to examine the dice, after which he demanded that Buffett choose first.

Nontransitive dice seem like a natural way to introduce nontransitive elements into a game. For example, in our battle game, Infantry could roll one type of dice, Cavalry another, and Artillery the third. This would give you the nontransitive effect for free – you wouldn't need any modifiers, special ratings, or anything like that. Just roll and go.

Make Your Own Efron Dice:

Die A:

0	0	4	4	4	4

Die B:

3	3	3	3	3	3

Die C:

2	2	2	2	6	6

Die D:

1	1	1	5	5	5

Each of these dice beats the next one in the list 67 per cent of the time.

Chapter 17

Making and Breaking the Rules

Analogies are the Foundation of Thought

In Chapter 4, I talked about complexity in games, and introducing new ideas. There I declared that to be easily digested, a game should have at most one new element or innovation. I'd like to expand on that a little bit here, in terms of writing your own rules to games.

First, there is some concern that the gaming public has a reduced attention span. And while I think there is some truth to that, I believe that actually we've been victims of our own success. Games are being played by a wider range of people, and the population of those that like super-complex games has not grown. I don't think there are fewer people in the world who like complexity. But as a percentage of the overall game-playing population? It's definitely decreased. Designers and publishers trying to capture market share need to take that into consideration.

In addition to limiting the number of truly new mechanics in a game, designers have another option: make the new mechanic analogous to something in the real world — something that most people already know about.

For example, let's say you're designing a game and have pieces moving on a board with squares. You have a piece and put a big '3' on it, and tell players it can move three spaces. Players may have questions: can it move fewer than three spaces? If I move fewer than three, can I move more next turn? Can I jump over enemy pieces?

But if I call that same piece a 'tank', I don't have to spell that out. By drawing an analogy with what they know about tanks, players will assume that you can't skip spaces, you can't jump over enemy pieces, and you can move fewer than three squares if you want.

If I want to make a piece that can move through the enemy, I can put a picture of a helicopter on it, and my job is pretty close to being done. If it has to move the full amount — or at least one square each turn — maybe I put a picture of a jet on the piece.

By selecting the right analogy, by fitting the theme and the mechanic together appropriately, you can dramatically reduce the complexity in a game.

When I'm teaching *Space Cadets: Dice Duel*, a game about battles between *Star Trek*-style ships, there's a moment where I explain how weapons work. I show people different faces of the dice and say, 'A torpedo is made up of three parts: this is the nose, this is the body, this is the tail. And you put them together like this, and you have a torpedo.' And I love watching people at that moment, because that just *makes sense* and their faces light up with understanding. I don't need to explain it ever again.

Similarly, we've all seen examples of game elements that just don't fit what they're called. If I design a game where airplanes can't move over tanks, that will make the game harder to learn and less engaging.

So-called 'gateway games' like *Dominion* get a bad rap. Some people treat them as a way to get people interested in games, putting their toes into the water before diving in. Others complain that they are dumbed down, not as deep, not worth wasting time on.

But here's the thing about gateway games. They introduce players to a single innovative mechanic in a controlled way. *Dominion* introduces the mechanic of 'deck building', where players start with weak cards and gradually add more powerful cards, with more and more synergies, to their decks. These are not just 'simple' versions of the 'real' games that we actually want to play. These are ways of giving people mastery of certain concepts that are shared among many games that are out there.

No one likes to have no idea what they are doing and drown in a sea of innovation. People like to feel intelligent. They like to feel in control. Through exposure to lots of different games, and repeated play, experienced game players are able to look at just about any game and form analogies that will guide them, but casual players are not. However, everyone has a wealth of practical, real-world experience to build on. As game designers and teachers, we can latch onto analogies to make players catch on quickly, and feel like they understand what is going on.

So, if you're a designer and you must have more than one innovation, make sure they are not tightly coupled, so that players can focus on one and still do okay, and make sure you have drawn a nice analogy from the game to something that players are already familiar with.

In their book, *Surfaces and Essences*, psychologist Emmanuel Sander and cognitive scientist Douglas Hofstadter claim that analogies are the foundation of all of our thinking.[1] They are not just a useful tool, but the absolute way that our brains work. And they are not talking about analogies you find on tests, like '"dog" is to "puppy" as "cat" is to "_____"?' They are talking about everything from words and phrases, to categories like 'sour grapes'. Making an analogy is taking something new and relating it to something you know about: putting it into categories. Recognising something as a 'chair' or 'the letter B' is similar to recognising something as a 'sour grapes' attitude.

Hofstadter and Sander take a very strong position on the power and role of analogies. They put analogies as the central factor (possibly the *only* factor) in cognition. Their definition of intelligence is that, when faced with a new situation, you are able to swiftly and accurately draw parallels with a precedent stored in your mind – to form an analogy between what is happening now and what happened before, even though at a surface level those two things seem very different.

As humans, we are hardwired to break things into groups, to label things. It's a way to make sense of a chaotic world: 'This is safe to eat; that will try to eat me.' The urge to categorise goes back to our earliest stories, with Adam naming the animals, through to the biologists who categorised all the plants and animals into Domain, Kingdom, Phylum, Class, Order, Family, Genus and Species. (Or, as the mnemonic device my kids came up with goes: Donkey Kong Plays Centipede Only For Gold Stars.)

Back in the 1870s, mathematician George Cantor developed 'set theory'. This was a way of manipulating groups of information. In set theory, everything is either *in* the set or *out of* the set. There is no grey area. This is related to Boolean logic, developed by George Boole about 20 years earlier, where everything is true or false.

However, the real world isn't that simple, so categorising and analogising items can sometimes get us into trouble.

Meet Shaquille O'Neal, whose height is 216 centimetres.

Now, let's take the statement, 'Shaquille is tall'.

Is this true? I would say yes. But where is the cut-off between tall and not-tall? Is 182 centimetres tall? If so, is someone not tall if they are 181 centimetres? That doesn't make any sense. There is no real cut-off between tall and not-tall. So, does this mean that the statement 'Shaquille is tall' really doesn't mean anything?

That's obviously not the case. We use words like that all the time, and we all know what we mean. So, can this type of language be formalised in some way?

Well, this was the question that confronted mathematician and computer scientist Lotfi Zadeh in the 1960s. He realised that one way to look at it was to allow for something to not be completely in or out of a set – but only part way in. He called his new idea 'fuzzy logic'.

In fuzzy logic, every item has a value of between 0 and 1 that says how much it is part of a set. For example, let's take 'tall'. Who is in the set 'tall'? Well, Shaquille, at 216 centimetres, may be a 0.9 (or 90 per cent tall). I'm 183 centimetres, so I may be a 0.5 (or 50 per cent tall). Someone who is 157 centimetres may be a 0.1 (or 10 per cent tall).

Zadeh determined ways that sets of these 'fuzzy' elements could be manipulated and combined, and it has been an interesting, albeit controversial, area of research since then. Some say that fuzzy logic isn't necessary – that regular set logic can accomplish everything that fuzzy logic can. And some say that binary logic is just a subset of fuzzy logic, and that fuzzy logic is the broader theory.

So, ironically, mathematicians are arguing over whether fuzzy logic should be in the set of 'useful theories' or not.

This has implications when it comes to games as well. For example, is *Memoir '44* a 'wargame'? Is *Gloomhaven* a 'Euro game'? Is *Dominion* a 'gateway game'?

Fuzzy logic gives us a way of looking at this problem. We might be able to say, for example, that *Memoir '44* is 40 per cent 'wargame', while *Advanced Squad Leader* is 90 per cent 'wargame'. Of course, everyone will have their own assignments. I would say that *Panzerblitz* is more of a wargame than the miniatures game *Flames of War* due to its more detailed combat resolution system, but I'm sure others would assign different 'wargaminess' to them, based on the elements they consider important (and analogies they use) in determining what a 'wargame' is. For example, I think it's important that if tactics would fail in reality they should also fail in the game. Others may think that having the right unit organisations and markings is important.

Next, the question of threshold comes in. I may call something that is 50 per cent 'wargame', a wargame. And someone else may say only 25 per cent is necessary.

We've got two different levels working here: what percentage you would assign to the game, and what your threshold value is. Both of these are totally subjective, and that's why arguments about categorising games go on without any resolution. It just isn't possible.

So, while there is value in trying to sharpen terms – to help us communicate with each other – trying to definitively corral games or any other objects into specific categories is doomed to failure.

Picking the Right Metaphor

A famous designer (who shall remain nameless) once told me, 'If one more person asks me if I start with mechanics or theme first, I'm going to punch them.' The reality is that mechanics and theme should work harmoniously together. In a very real sense, one can't exist without the other.

Robert McKee, in his book about screenwriting, *Story*, says that he is constantly asked which is more important in a story: plot or character. And he says that the question is meaningless. Plot and character are, in essence, the same thing. The plot arises from a character making decisions. And those decisions define the character. The plot and characters feed into each other in a loop – characters making decisions that move the plot forward, and that plot movement forcing characters to make decisions on how to react. And around and around it goes.

In a game, the connection between mechanics and theme might not be quite as tight, but when they are at odds with each other, it can undermine what the game is trying to accomplish. The combination of the two results in the player experience, and it is the *experience* that good designers focus on.

Let's do a thought experiment. The wonderful *Incan Gold*, or *Diamant* as it is also known, by designers Bruno Faidutti and Alan Moon, casts the players as Indiana Jones types, exploring deeper and deeper into a cave to try to grab the most gems. Each turn, each player makes a simple decision – move further into the cave and risk losing all their loot, or go home and bank what they've found. As the cave fills with hazards, the pressure builds.

The mechanics and theme are perfectly intertwined in a way that supports the tension and the narrative.

Flash Point: Fire Rescue, from designer Kevin Lanzing, is a cooperative game about firefighting. The players must work together to rescue at least seven out of ten people trapped in a burning building. Like *Incan Gold*, it also features push-your-luck mechanics, as the players have to take chances in guessing how the fire will spread.

Flash Point is another excellent integration of theme and mechanism. The cooperative nature and the spreading of the fire all work together to again build tension and give the players a satisfying experience.

So, let's combine these games. Let's take *Incan Gold* and retheme it as firefighters trying to rescue as many people as possible from a burning house before it collapses.

So now, unlike *Flash Point*, this is a competitive game. But players don't directly get in each other's way – it's as if each is pushing forward into the building separately and trying to rescue people. Or we can say that the farther they get into the building, the more 'heroism' points they earn.

When the firefighters think that it's too risky to continue to push on, they leave the building, and collect no more heroism points. Other players may try to earn more points by staying in the building longer before leaving.

Mechanically, it is exactly the same as *Incan Gold*. But would you be as satisfied playing this game? As a player, you would have a different experience. In *Incan Gold*, players are Indiana Jones-type adventurers trying to push their luck to get gold and jewels, and if they stay in too long and get crushed by a boulder, other players can point out that they were too greedy. But if players are firefighters, and they push too far into a burning building to try to save more people? It would be a tougher decision to turn back and leave the building, even at the risk of losing the game. Psychologically, the new theme turns the decision it wants the players to make on its head. In *Incan Gold*, leaving and not being greedy is the safe, sane and society-approved choice. In our fire game, however, leaving early is cowardly, against the code of firefighting and putting other people's lives at risk. The theme and mechanics combine to make a very different experience for the players, one that may differ greatly from what the designer intended.[2]

I experienced a similar thing when designing *Survive: Space Attack*. In the original *Survive*, the players are attacked by other players in the guise of sharks and sea monsters. In our initial retheme, we replaced the sea creatures with 'rebel fighters' that attacked the players' tokens as they tried to float to safety in escape pods. But there's a real difference in attitude when you use a shark to go after someone swimming than if you're piloting an X-wing and shooting someone floating in a space suit. Yes, they both end up dead, but the psychology of it is completely different. We ended up scrapping that whole design for generic bug-like aliens, which fulfilled the same roles.

As a designer, you need to pay attention to the linkage of theme and mechanics, and make sure that they reinforce where you want to take the players.

Power

One of the fun things about playing games is that they give us a chance for wish fulfilment – to live out fantasies of doing things we couldn't or wouldn't do in real life. And for many folks, one of those fantasies is to be powerful.

Many people feel or have felt powerless, and the opportunity to have power drives a lot of the entertainment we enjoy. Pretending to be superheroes, wizards and Jedi are all examples of this desire for the powerless to play at having power.

Mundane games often don't have the spark that wielding power gives, and many people decry them because of this. Take *Thurn and Taxis*, for example. Maybe it's mechanically sound, but who wants to pretend to set up German postal routes when you can throw a fireball at your opponent?

So, what is power, and how does it manifest itself?

In 2011, a team led by psychologist Gerben van Kleef of the University of Amsterdam did a study that attempted to shed light on some of these questions.[3] They started with the hypothesis that people perceive powerful people as those who don't necessarily follow the rules. In other words, power corrupts.

They did several experiments to explore this hypothesis. In the first, they had subjects read stories about people and rate them for, among other things, how powerful they thought the characters were. In stories where a character broke rules – by taking someone else's cup of coffee without asking, or putting their feet up on a table in a cafe, or dropping cigarette ash on the floor – subjects thought that those characters were more powerful.

The same thing carried over in face-to-face interactions. This time, subjects were split into two groups. In both groups, experimenters had the subject sit in a room, and an actor came in and interacted with the subject. In one group, the actor was polite and followed the rules. In the other group, the actor was obnoxious and broke rules. For example, they would come in late, throw a bag heavily down onto the table and then put their feet up on the table. After the interaction, people thought that the rule-breaker 'had more power and was more likely to get others to do what he wants'.

So, when people break rules, others think they are more powerful than those who follow the rules.

Let's apply this to game design.

For players to have more fun, we want to make them feel powerful.

People think that people who are powerful break rules. So, we need to give players the ability to break rules.

Let's take a look at *Cosmic Encounter*. As I mentioned earlier, the mechanics of this game are dirt simple. They are basically just a variation on the card game *War*, with players able to negotiate, then fight to dominate or settle their dispute peacefully. However, what makes *Cosmic Encounter* so successful (first published in 1977, it has been reprinted countless times) is that each player gets something called an 'Alien Power Card'.

Here is an excerpt from the rules of the Mayfair version, published in 1991:

Each player takes the role of an alien race determined
to gain control of the universe. The Alien Power Card

describes the unique power that each race possesses. Players can use the unique power of their race to break certain rules in the game.

Wow! That hits all the buttons right there. The cards themselves are called *Power* cards. They are giving powers to the players. And it specifically says that your power is used to *break* the rules of the game.

So, here, the designers of *Cosmic Encounter* are slyly using the psychology of the players to make them feel more powerful. The alien powers are part of the larger rules of the game. You're not breaking the rules willy-nilly – you are 'breaking' rules in a way that is specifically designed by the game designer.

Another key feature of the *Cosmic Encounter* design is that each player gets to break the rules in a way that only they can. If everyone could break the rules in exactly the same way, it wouldn't have nearly the same psychological impact. For example, in the game *Finca*, each player gets a few special-ability tiles that they can choose to play during the game. But since all players have the same abilities, it doesn't give that same sense of 'going rogue' and really breaking rules.

This is also underlined in *Cosmic Encounter* by the text on the Alien Power Cards themselves. They all start with something like 'You have the power to Multiply' or 'You have the power of Knowledge'. This further personalises your ability and emphasises that you have a particular role to play in the game.

Giving specific unique identities to players also adds other psychological aspects to the game, like jealousy, that further accentuate the power that players can exert over each other.

There are tons of games that use the idea of variable player powers to feed into the player's desires for strength, mastery and escape from the real world, if only for an hour. A commitment to the small details of how this is implemented – like specifically calling things 'powers' and telling players that they are 'breaking the rules' – can make a huge difference in immersion and the success or failure of the game.

Chapter 18

Community of Players

Microbes, Communities and You

It turns out that we are mostly not us. And by 'we', I mean the cells that we carry around as part of this thing we call our 'body'. There are about ten times more microbial cells in our bodies than cells that have our DNA.

Advances in rapid DNA screening and sequencing have given scientists an unprecedented look into the micro-environment that is each of us, especially all the various critters that live in our digestive tract. Eating different types of foods can vary the proportion and composition of these microbes. They help us in innumerable ways to process and extract energy from food, and protect us from harmful effects. But if they get out of whack, it can contribute to obesity, Crohn's Disease and other maladies. New treatments that focus on restoring the correct balance of these microbes have proved very effective in early trials.

Interestingly, what these colonies inside us are composed of varies in different regions and cultures in reaction to the different types of foods and environmental effects they are exposed to – in fact, these colonies are so unique that they can almost be used as a fingerprint of sorts. A lot of the stomach problems that people have when travelling are related to their microbes being out of balance with the foods and environment.

Remember that these passengers in our bodies outnumber the real cells by ten to one. So, in one sense, we are hosts to this vast colony in a symbiotic relationship.

And this is very similar to the way that a game and its players exist in a symbiotic relationship. The game is like our body, and we the players – well, we are single-celled organisms that live in an intestine.

Bear with me here.

The game provides the framework within which the players 'live'. And, like the microbes, we can either help the game to live a long, healthy life. Or we can make it sick and even die.

I like this analogy, as it puts an interesting spin on player–game dynamics. Game design is a creative endeavour, and arguably an artistic one as well. Certainly, it is as much of an art form as a movie or novel. All forms of art depend on the interaction between the object and the viewers. However, this interaction is the lifeblood of a game. And interactions are not just between an individual player and the game, but also between the players themselves. Games, much more so than other artistic forms, are brought to life by how the players approach them.

In the 1981 solitaire game *B-17: Queen of the Skies*, the player leads a crew through a series of missions over Germany during World War II. The game doesn't really have that many decisions. You primarily roll on a series of tables to see what happens as you progress through your mission.

And yet, *B-17* has enduring popularity, even 30 years later. There are organised play groups, and people who track statistics for all their crew members and get unbelievably attached to what happens to these imaginary people.

But if you don't make that imaginative leap – if you peel back the layer over the paper-thin mechanics – there's just a glorified accounting exercise underneath. I can totally understand someone being completely bored by *B-17*. And yet there are plenty of players who bring passion to the game, who weave those dice rolls into elaborate narratives that continue to enthral them for years. They are being good little microbes.

As a game designer, you want to make sure that you attract the right kind of players to your game. If there is a mismatch between the players that buy your game and who the game is designed for, your game will sicken and die. And it's not just about whether your mechanics have 'just enough' balance of choice versus chaos, as we discussed in Chapter 15; or 'just enough' innovation for a gateway/non-gateway game, as was the case in Chapters 4 or 17. The theme and graphics also have to attract the type of player who will find the mechanics enjoyable.

For example, the game *Krosmaster: Arena* is graphically very cute and accessible, with chibi figurines and 3D terrain. Because of the graphics and components, it is going to attract younger and more casual players. The success of *Krosmaster* can be attributed in large part to the simple mechanics that appeal to those types of players. If those components were grafted onto a more complex rules base, the game would have baffled and disappointed a lot of the casual gamers who gave it a shot.

On the other end of the scale, a wargame like *A Distant Plain* promises to be an in-depth examination of the ongoing conflict in Afghanistan. That is going to attract a certain type of player who is looking for fidelity to the subject matter. And again, there is a good match between expectations and reality.

On the negative side, I will point to my own *Space Cadets*. The subject matter, cartoony graphics and technobabble promise a quick, whimsical game. And while most of it is simple, there are enough fiddly bits to the rules that, for some players, it doesn't meet that expectation. But with the right players – the right population of microbes – it really comes alive and shines.

On this topic, let's talk about long games.

Long games certainly have an advantage in fostering immersion. If you're playing something for four hours, you can get deeply involved in the unfolding action. At the same time, they can be a nightmarish death march if not designed properly, or played with the wrong microbes/players or under the wrong circumstances.

There are a few key factors for making a long game work. The first is engagement – how much attention is required of the players. The two best options here are, oddly, the complete opposite of each other.

The first option is constant or near-constant involvement. For example, a game of *Diplomacy* can easily last six hours and players are constantly engaged throughout. There are three basic things that players do: talk to each other, write down the orders for their units and then execute the orders.

When you're negotiating, you are completely engaged – perhaps even more engaged than you usually are during a conversation, since you have to figure out who, if anyone, is telling the truth.

When you're writing your orders down, it only takes a couple of minutes, but you need to make that final decision about what to do and make sure you're writing things down correctly. And in the execution phase, you have the tension of seeing what the other players do and quickly trying to explain why you didn't do exactly what you said you would.

Now each of these phases requires 100 per cent of the players' attention, and they all happen simultaneously. There is no downtime in a game of *Diplomacy*, and you really don't feel the passage of time. You are constantly engaged.

Advanced Squad Leader doesn't have to be a long game depending on the scenario you're playing, but it can be. And the way it keeps players involved is by the mechanic of interrupting the other player. As one player moves their forces, the other has the option of stopping them at any point and shooting them in their current spot. You need to be aware of what the other player is doing, try to figure out their plan, and see what you can do to disrupt it.

Then there are games that are at the other end of the scale, where a player turn takes a long time and the other players are not engaged at all during this time. In longer more complex games, like *Warhammer* or *Warmachine*, your opponents' turns can easily be 30 minutes to an hour. You don't see games like this much any more, but there were plenty that used this model back in the 1970s and 1980s.

Surprisingly, though, they worked. And the reason they worked is because the other player could walk away. You knew that when your turn was done, you had 20 minutes to kill while your opponent made their move, and you could go off and look at another game or a magazine, watch TV, chit chat with someone else, whatever. You didn't need to stay involved for that time so you could turn your attention elsewhere.

The problem games are those that lie in the middle. For example, a game where you're just waiting for your turn to come around, but there may be a quick step in between that needs your attention. Or the time between turns may be long, but not long enough that you can't simply check out of the game.

That's the toughest type of attention to give to a game — a low level of constant attention with punctuated points of total focus. And it's hard to maintain that level over the full course of a game. The game *Risk* is that way to a certain extent. You really can't leave the table in case someone attacks you. But there may be long stretches where you aren't doing anything. And even shorter games can have this problem, as I'm sure we've all experienced.

Now, there have been a number of studies about attention spans, multi-tasking, and related topics over the last 20 years. But there are many conflicting results. Some say that our attention span is being eroded because of the internet and the endless source of distraction it supplies. Others say that our attention span is the same as it has ever been. Some say that video games cause distraction; others that video games can help you keep focus for extended periods of time.

So, I'm not sure where the truth lies.

However, one thing that most agree on is that focus and attention are a muscle you can exercise. The more you practise staying on task, the easier it will get.

And games are certainly a good way to train that muscle.

Gamification of Life

While writing this chapter, I stopped for lunch at my local bakery, and won a free smoothie through their new loyalty program (I opted for strawberry). Normally I don't sign up for those programs, but I usually eat once a week at this bakery and, more importantly, they've designed their loyalty card to be a game.

It's not a particularly challenging game, but it's still a game. Whereas most programs stamp a card, and give you a discount or free meal or whatever after a certain number of visits (with or without a few free stamps to start you on your endowed journey), this bakery's card might randomly give a bonus each time I visit. It's like a slot machine. Most of the time you get nothing, but sometimes you get a free smoothie, or something even better.

This is an example of games leaking into everyday life. And this has been a fertile area for development in the last year or so. It seems like everyone is tacking some sort of game mechanic onto their product or service.

For example, the social-media app Foursquare uses the GPS in your phone and lets you post about where you are, see if your friends are nearby, and read people's tips about local businesses. Foursquare is solidly based on the concept of 'badges'. These are earned by doing different things — checking into lots of different cities or locations, for example. In October 2010, astronaut Doug Wheelock was the first to unlock the 'NASA Explorer' badge by checking into Foursquare from the International Space Station!

In addition to earning badges, users can also become the 'mayor' of a location by checking in the most times in a 60-day period. In addition to bragging rights, certain businesses offer discounts or other offers to the current mayor.

Both of these concepts are taken from video games. Achievements go back at least to the original Xbox in 2001. Certainly, the concept of 'experience points' goes way back to the original *Dungeons & Dragons* in 1974. These are powerful concepts that tie directly into the reward centre of our brain. We will do things to get stuff. Even pretend stuff. The success of *World of Warcraft* is due to this effect. People like to watch their experience bar fill up and to work their way toward that next cool new power, or battle the same monster for the hundredth time hoping that this time the magic slot machine will drop something epic.

There have been some attempts to harness this wiring for good. When you first start up the phone productivity app *Epic Win*, you pick a character: Dwarf, Warrior or Treeman. You then load your to-do list and rank the tasks for difficulty. When you mark the items as done, you gain things – loot, progress in a quest, gold and experience points. As you level up, you will unlock more quests, which will require you to complete more tasks. And so on and so on, until eventually you have replaced all the light bulbs in your house, weeded the garden and finished that remodelling project.

Similarly, I now have a baked-goods slot machine in my wallet. And the bakery is smart to make it a slot machine. Random rewards are much more addictive than rewards that occur at regular intervals. Psychological studies have shown this time and time again. And certainly, the success of *Magic: The Gathering* is due to their slot-machine sales model – although the concept of Fantasy Flight Games's 'living' card game or other companies' 'expandable' card games are testing this idea. A living card game sells additional cards in pre-defined packs. You know exactly what cards you get, unlike in a game like *Magic*, where the cards you get in a pack are almost totally random.

I have a card that gets stamped each time I go into my friendly local game store and buy something. And I know that after I spend a certain amount of money, I will get five dollars off my next purchase. It's nice, but I don't think it really makes me spend any more money.

But what if every time I made a purchase, I got to spin a big wheel that might give me five dollars off, or might give me a $50 coupon? That would be cool. The store could set up the odds so that on average they are giving the same discount as with their current loyalty card, but it would be so much more fun, and actually get people to buy something just to get to spin the wheel. Because – hey, you never know.

Colonoscopies and Board Games

No, that title is not a joke. This is real science. And sometimes science is not pretty.

Scientists were interested in learning about how we remember experiences. When we look back at something (say, a colonoscopy ... or a board game), how painful or pleasurable was it compared to other experiences?

In the 1990s, physician Don Redelmeier and Daniel Kahneman did a study at the University of Toronto.[1] Working with people undergoing colonoscopies, they asked them to record, at 60-second intervals, how much discomfort they were experiencing, on a scale from 1 to 10.

Let's look at two representative patients: Patient A and Patient B. Patient A's procedure lasts for ten minutes. He records the worst pain at about an 8, and it happens close to the end of the procedure at the nine-minute mark.

Patient B also records the worst pain at the nine-minute mark, but his procedure lasts for 20 minutes, not ten. The first ten minutes recorded by A and B are almost exactly the same, spiking up to 7 or 8. But B has an extra ten minutes, during which he records the pain at around the 4 or 5 level.

So, which would you rather be? It seems like Patient A gets the better deal here. Both A and B experience the same thing over the first ten minutes, but then A is done. B has an extra ten minutes before he is finished.

Well, after the procedure was over, the experimenter asked the patients to rate 'the total amount of pain' they experienced. Patients that experienced procedures like Patient B reported much less total pain than Patient A.

Psychology researchers prior to this always assumed that the 'total pain' of an event would be the sum of all the individual pain recordings that people made during the procedure. This experiment, and many others done since, prove otherwise.

It turns out that people assign a 'total pain' value based on the average of two things: the worst pain and the pain at the end. The duration of the pain has no effect on how we remember things. So, Patient A, who had his worst pain right at the end of the procedure, has a much worse memory than Patient B, who had less pain right at the end – even though his procedure was twice as long.

There are two 'selves' that we have: the 'experiencing self', which is what we are actually going through at the time, and the 'remembering self', which is how we remember events later.

This raises an interesting ethical question: would you rather experience more pain but have a better memory of the event? Or experience less pain, but remember it worse? This is not an idle question – it is something being explored today.

What does this have to do with board games?

Well, it turns out that the same rules apply to things you enjoy — not just pain. The memory of an enjoyable event is based on the peak enjoyment, averaged with your enjoyment at the end.

And this is why the end of a game is so critical. We've all had the experience of having a good time during the game, but then the end just falls flat. And we remember it as a 'bad experience'. Or we play a game that's not so exciting, but suddenly there's a dramatic twist right at the end and we remember it fondly. If that big twist happens in the middle, it is not nearly as memorable.

And looking at it logically, that really doesn't make much sense. If you are enjoying yourself for two hours, but are unhappy with ten minutes out of that, haven't you had fun overall? Why does it matter if that not-as-exciting ten minutes happens during the middle or at the end? It shouldn't. But it does. Our brains encapsulate events by looking just at the peak and at the end.

So, game designers take note: endings matter. You knew that intuitively, but now you know it scientifically. People are truly their 'remembering selves', not their 'experiencing selves', and we need to design experiences with that in mind.

Chapter 19

Emergent Properties and Games

Emergence

Whenever I take a long trip, I like to read game rulebooks.

Yes, I admit it, I like to read rules. Just for fun. I do it for a variety of reasons – to see if it's a game I would be interested in purchasing, to refresh my memory on an old game I've forgotten the details of, or to see if there are any innovative ideas contained in the latest offerings. Sometimes I feel like a rules vampire: I swoop down on a new game, digest the rules, suck them of anything interesting, and then move on to the next target.

I own a stupid number of games (1949 at the time of writing), which means I've read through a lot of rules. And what I look for before adding a game to my collection is a difference in theme or mechanic or artwork or ... something. I often think, 'Hmm ... pulled this mechanic from this game, this mechanic from over here, stirred them up, pulled the theme from this computer game, and presto … it's a new game.' It's hard to come up with really novel mechanics — especially ones that work.

Now, interestingly, when I started gaming, I focused on theme. I bought wargames or fantasy games because they looked cool. Then in the 1990s, as we started playing more Euro games, I developed a deep appreciation for mechanics, and that alone was enough to carry a game for me. But now I find myself swinging back to the theme side — or perhaps it's better to say that I prefer a strong theme melded with strong mechanics. Mechanics alone are usually not enough to make me want to buy a game.

So, here's my question: can you understand a game by reading the rules?

Now, at first glance, there seems to be an obvious answer. Yes, of course. You must be able to, because the rules, by definition, are the game. For certain games, we may have to extend the definition to include the components, because sometimes rules are there as well — like the Alien Power Cards in *Cosmic Encounter*.

However, I would maintain that there are features in a game that you cannot anticipate unless you actually play the game. Subtle interactions, pacing, strong and weak plays, and overall strategies cannot necessarily be predicted or understood just by reading through the rules and looking at the components.

Games exhibit what are called 'emergent properties'. This means that there are features of the game that are the result of the interactions of different parts of the game that would not be predicted just by looking at the separate parts themselves. The whole is, in this case, more than the sum of its parts.

As a classic example, let's take a look at the ancient game *Go*. In *Go*, two players (white and black) alternate placing stones on a board. Once placed, the stones do not move. If two or more stones of the same colour are connected orthogonally, they form a group. A group is captured if every space adjacent to it is occupied by an enemy piece. You are allowed to place a stone into a position where it would normally be captured if, in the act of placing, you capture enemy stones. At the end of the game, you get one point for each empty space of territory you have surrounded, and one point for each enemy stone you have captured.

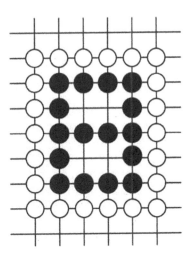

In this Go game, the black stones have two eyes.

That's it. Those are pretty much all the rules. But from those very simple rules, a tremendous number of complex emergent behaviours and patterns arise. For example, certain configurations of stones cannot be captured. These groups share the feature that they have two separate empty areas, called 'eyes', inside them. So, an important part of the game is making sure that you give your groups sufficient eye space. From the fight for territory and eye space, it turns out that the best strategy at the opening is to begin by trying to grab the corners of the board, then to move up the sides, and finally to go toward the centre.

Both of these features of *Go* – the fact that groups with two eyes cannot be killed, and the superiority of starting in the corners – are not at all obvious the first time you sit down to play, and probably aren't even on your radar. You need to start putting pieces down on the board and play several games before these patterns become apparent. I am pretty sure if you gave someone the rules and asked them to imagine how a game would progress, or whether the game of *Go* would be deep or fun, people would be hard-pressed to jump from the rules to what gameplay is actually like.

The quest for fundamental laws in science creates a similar effect. Science is effectively like looking at games backward – we watch people and things play and try to figure out the rules. And then, once we learn the rules, we use them to build up predictions about what will and will not happen. But because of the phenomenon of emergent behaviour, it is ridiculously difficult to start with the core set of rules – the basic scientific laws we all know – and predict complex behaviour.

This would be like starting with a theory of gases and heat and pressure, and trying to predict that hurricanes would exist. Or starting with quantum mechanics and trying to predict that superconductivity would be possible. Or that termites, following simple rules, will be able to build amazingly complex structures. Remember our NP Complete problems from Chapter 14? Each of those had very simple rules, and very simple goals, and yet 'solving' them has proved mostly impossible.

It is easier to look at what's out there and deduce basic principles than it is to take the basic principles and extrapolate everything that is possible in this complex, wonderful world.

So, from that perspective, just reading rules on a long trip really isn't a fair way to judge whether a game will be good or not, or to really understand what the designer was shooting for. That's why designers need to playtest their games with other people as much as possible. That is what will allow those hidden properties to be discovered.

But deep down, I guess I just enjoy reading rules, so I'll keep on doing it.

Predicting Complex Behaviour

Many emergent properties in games begin with the players. To illustrate this, I'll start with a quote from the excellent movie *The Princess Bride*:

Westley: All right. Where's the poison? The battle of wits has begun. It ends when you decide and we both drink. And find out who is right — and who is dead.

Vizzini: But it's so simple! All I have to do is divine from what I know of you. Are you the sort of man who would put the poison into his own goblet or his enemy's? Now, a clever man would put the poison into his own goblet, because he would know that only a great fool would reach for what he is given. I am not a great fool, so I can clearly not choose the wine in front of you. But you must have known I was not a great fool – you would have counted on it! So, I can clearly not choose the wine in front of me!

Westley: You've made your decision then?

Vizzini: Not remotely!

We've all been in this situation in games. Did your opponent do the most obvious move? Or did they do the not-obvious, figuring you would guess they did the obvious? This feature is in just about every game with simultaneous or hidden decision-making, from *Citadels* to *Diplomacy*, *Poker*, and *Ys*. And it tends to bring out strong emotions in people as they try to outthink their opponents.

So, does science have anything to say about this? Is there a way to predict what your opponent will do? Experiments have been done on these questions, and the answers are interesting. I'm not sure how much they're going to help you in your next game but let's take a look anyway.

In a 2005 contest held by the Dutch newspaper *Politiken*, participants were asked to pick a number between zero and 100. All the numbers chosen would then be averaged, and the winner would be the person whose number was closest to two-thirds of the average. Not the average itself, but two-thirds of the average. That's the crux.

These types of problems, and there are several variants, are called 'beauty contest' problems, because they are measuring the popularity of certain choices.

Now, before we move on, take a second and think about what your entry into this contest would be.

And we're back. We've talked about the concept of the Nash Equilibrium in other chapters. Basically, it's a strategy that everyone will use if they're all super logical. And this game has a Nash Equilibrium: picking zero. If everyone picks zero, everyone wins – because two-thirds of zero is zero. And if you go higher than zero, and most everyone else picks zero, the two-thirds average will still be close to zero. So, deviating doesn't help you.

So logically, you should have picked zero.

People, of course, are different. Very few pick zero. The actual results, in fact, show peaks at characteristic values. The biggest peak is at 37, the next biggest is 50, and then 25. Only about 2 per cent of people pick zero.

The beauty of the way this game is set up is that you can see how many steps people go through before coming to their answer. In the *Princess Bride* problem, it's just two outcomes – either Vizzini picks the poison and is dead, or he gives Westley the poison and remains alive – so there's no way to see if someone goes two steps deep or four steps deep. But in this problem, you can tell.

Let's go back to the common answers.

First, let's look at the 15 per cent that said 50. These are the math-challenged folks who saw the word 'average' and picked 50 without really thinking about the answer. In addition, there is background noise in the data, with a consistent number of people choosing non-peak values (all the way up to 95, surprisingly) that appear to be random choices. So, if you add the people who picked 50 to the people who picked random numbers, you get about 50 per cent of people falling into this category.

The largest peak, with about 25 per cent, is 37. These are the people who figured out that the average player would probably pick 50 or just randomly, so the average should be around 50, and went to two-thirds of that value, which is 37.

Next up is 25, selected by about 15 per cent of responders. These people went one more step, figured most answers would be around 37, and went to two-thirds of that value, which is 25.

Scientists have developed a model for this called the 'cognitive hierarchy' model.[1] Basically, it says that there are certain levels of players that base their answer on the levels below them. So, level 0 players pick either randomly or the obvious choice, where level 1 players assume that the other players are all level 0, and act accordingly. Level 2 players take into account the existence of a mix of level 0 and level 1 players, and plan accordingly.

Now, modelling the percentage of players across these levels by the Poisson distribution (named in honour of French mathematician Simeon-Denis Poisson) actually gives a pretty good fit to this data across several different types of experimental games. A good approximation is that 50 per cent are level 0, and the number is cut in half each step down. So, 25 per cent are level 1, 12 per cent are level 2, and so on.

Very few people go more than three steps deep in their analysis.

Now, obviously, the distribution of people will depend on the conditions of the game and the players. For example, if you sit down in a room full of mathematicians and give them 15 minutes to think about their answer, you're going to get a different response from if you just stop people on the street and ask them for an instant reply.

So how does this help us play? Well, I don't know. If you're in a situation where level 0 and level 2 thinkers come up with one move, and level 1 and level 3 thinkers come up with the opposite, you can figure that two-thirds of the time the basic level 0 choice will be made, and one-third of the time the

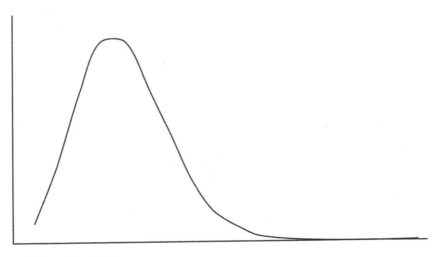

A Poisson Distribution

opposite choice. But it's a statistical thing in the end, and very dependent on the player's personality, familiarity with the game, mathematical sophistication – all factors that have to be taken into consideration.

The final answer to the contest, by the way, was 21.6.

The Impact of Added Players

We have already seen in other chapters how the impact of a third player can change the dynamics of a game. For instance, in the *Third-party Punishment Game* or adding more players to the *Ultimatum Game*, as discussed in Chapter 12, the third player adds a competitive aspect to the game that did not previously exist.

In much the same way, the addition of a third player in a tiered-power game can completely change the existing power dynamics and the odds of each player winning.

Say you are playing a game where each player begins with ten points, and can either remove two points from another player, or give one point to themselves. In a two-player game, one of those actions is more powerful than the other: remove two points from the second player and you are both hurting them and helping yourself (as you are now two points ahead). On their turn, they will remove two points from you, hurting you and helping themselves (bringing you both to a draw).

But in a three-player game, hurting the second player doesn't just help you. If you remove two points from the second player, both you and the third player are two points ahead of them. If the second player retaliates and removes two points from you, they are hurting you, helping themselves, and the third player is unaffected.

In both cases, you and the second player are tied after their retaliation, but the addition of the third player means someone else has now benefited from your actions.

If the third player takes the weaker action and gives themselves one point, they incur the wrath of no other player and still benefit from the mutually assured destruction between you and the second player.

Of course, this means in the next round, the third player is three points ahead, which makes them a potential target, so you can see how three-player games can have more to them than a simple tit-for-tat.

Let's play another game to illustrate this.

I've invited a cowgirl and a cowboy over for breakfast. The problem is, they brought their pistols and dawn is fast approaching. And cowboys and girls always duel at dawn.

Our cowgirl is Syd the Kid. Yee haw! And our cowboy ... well, he doesn't have a name, so let's just call him the Nameless One.

So, these two are going to have a gunfight. But it's going to be a polite gunfight, where they flip a coin and see who gets to go first, and then they're going to shoot at each other. And the last person left standing is the winner.

In this game, the Nameless One is the premier gunfighter in the county, and he has a 100 per cent chance of hitting whatever he's shooting at. Whereas Syd the Kid is a little bit less skilled, and only has a 75 per cent chance of hitting her target.

If we flip a coin and the Nameless One comes up first, he's going to shoot at Syd the Kid, and he's going to have a 100 per cent chance of taking her out. If Syd the Kid comes up first though, she's going to shoot at the Nameless One, with a 75 per cent chance of hitting and winning.

If she misses, the Nameless One has a 100 per cent chance of winning in retaliation.

So, Syd the Kid has a 75 per cent chance of winning (or $\frac{3}{4}$) 50 per cent of the time (or $\frac{1}{2}$). If you recall Chapter 5 on probabilities, because Syd must win the coin toss *and then* the shoot-out, we multiply the probabilities together: $\frac{3}{4} \times \frac{1}{2} = \frac{3}{8}$, or 37 per cent. Flipping that probability means the Nameless One has a 63 per cent chance of winning.

Let's mix it up a little bit, as our second cowboy arrives late for breakfast, but just before the sun crests the horizon. Now we have three in the gunfight: Syd the Kid, the Nameless One, and our third, One-legged McGee. This time, they're going to roll a die to see who gets to shoot first.

Now McGee, because of his one leg, is not quite as skilled as the other cowboys — so he only has a 50 per cent chance of shooting the person he's going after. This is where it's going to get a little more strategic and a little more interesting.

We're playing this game strictly logically. There are no grudges, and no one did anything to anyone's daddy or anything.

If the Nameless One wins the dice-off, and gets the first shot in the three-way gunfight, he is going to shoot at Syd the

Kid, because she has a higher chance of hitting him back if she gets to shoot.

In that case, he would take out Syd the Kid, then McGee would get to go next and have a 50 per cent chance of shooting back. Which gives both McGee and the Nameless One each a 50 per cent chance of winning that gunfight.

What if Syd the Kid shoots first?

If she shoots at McGee first and hits, then she knows that the Nameless One will take her out. If she shoots at the Nameless One and takes him out, then McGee is only going to hit back 50 per cent of the time. If she misses the Nameless One, then he will also shoot back and take her out. So, she's going to shoot at the person who has the 100 per cent chance of hitting her back. That part of the gunfight is the same as the two-player game, where Syd the Kid has a 37 per cent chance of winning. But either way, whoever comes out on top, McGee has a 50 per cent chance of shooting *them* and winning.

There's one more possibility we need to look at, and that's with McGee.

If he gets to go first, what is his best strategy?

You would think he should shoot for the guy who's sure to hit him: the Nameless One. Well, if he goes after the Nameless One and misses, then the Nameless One (tit-for-tat notwithstanding) is still better off going after Syd. But if McGee hits the Nameless One, then Syd the Kid is going to turn around and shoot at him, with a 75 per cent chance of winning.

Now, it seems like going after the Nameless One *would* be the best strategy, but it was actually a little bit of a trick question.

McGee's best strategy in this situation is … not to play. To shoot his gun straight up in the air and not aim at anybody. After all, non-violence is always an option … and letting the others shoot first gives him the highest odds of winning.

I tell this example because it illustrates a simple situation with the two players, where the player with a higher chance of hitting has a higher chance of winning the game. But as soon as we introduce that third player — even though that third player has a worse skill than the other two players — if you sit down and work through the math, it turns out that, assuming everyone plays their best strategy, then poor One-legged McGee, with only a 50 per cent chance of hitting something, actually has the best chance of winning the duel. In fact, he has a 54 per cent chance of winning, whereas the Nameless One is in second place with a 32 per cent chance, and the second-best shot, Syd the Kid, only has a 14 per cent chance of walking away from the gunfight alive.

Adding a third player into this mix can give you an extremely different experience from the two-player version.

Epilogue

Pivot Points and Phase Transitions

Reflecting on Phase Transitions in Life

Every day, we face lots of different boundaries – the transition between waking and sleeping, work and vacation, wanting and having. These boundaries shape and influence our lives, sometimes dramatically, but the effects are often subtle. Games are similar, and have different types of boundaries that impact play.

A boundary that all games have is their conclusion. Those that have a fixed number of turns, like *Terra Mystica*, often feel quite different on the last turn from the way they do the rest of the time. The value of points and actions becomes much simpler to calculate, and so, for many games, the last turn becomes an exercise in optimisation, as each player tries to maximise their victory points relative to the other players.

This is usually a bad thing in a design. It slows the game down, and can make the machinery of the game more evident — moves that otherwise are really bad suddenly become really good. So, designers either have to live with the consequences or patch the rules to disallow those moves, or otherwise inhibit the players.

In *Through the Ages*, one of the key mechanics of the military system allows players to sacrifice units to double their strength. Throughout almost the entire game, this leaves you more open to counterattack by the other players, so it is used sparingly. However, on the last turn, why not just sacrifice every single unit for a big war and lots of victory points? To counteract this, the designer added a rule that you are *not* allowed to sacrifice on the last round. This is just another rule that needs to be learned by the players and reduces the elegance of the system.

I call these 'edge effects'. As a designer, I always look out for edge cases, which can take a number of different forms. They can be the first or the last turn, or they can be people exclusively acquiring a single resource instead of spreading out. It is important to look for how edge effects can break a game.

Problems with edge effects are not limited to games. They can be seen in such diverse arenas as politics, where term limits can create a 'lame duck' president, whose ability to act is impacted, both positively and negatively, by the election of their successor, and computer programming, where edge effects account for the highest percentage of software bugs.

Like in politics, edge effects can also have a beneficial impact on a game design. For example, they can naturally create a rising tension and narrative arc to a game. Some games, like *Power Grid* and *1830*, add specific phases – mini boundaries that impact the flow around them. Anticipating and manipulating the phase boundaries in those games is a key part of player strategies.

Some games have more fluid, fuzzy boundaries. Two examples of this are *St. Petersburg* and *Dominion*. Both of these games have a pivot point – an inflection point – where the players need to transition between building up their engines and collecting victory points. You have to sense when the end is approaching and make that pivot at a key time to maximise your chances of victory.

Many people mark the milestones of their lives by decades. When you hit 30, 40 or 50, you sit back and take stock of where you are. I do things a little bit differently, which I'm sure surprises no one. When I was 16, I decided that the time for reflection would be the perfect square years: 16, 25, 36, 49 and, in a few years, 64. As it feels like the passage of time gets faster as I get older, I think that spacing these milestones further apart has worked pretty well. It also has tracked a lot of key moments in my life. When I did this for the first time at age 16, my parents were finalising their divorce. At 25, I got engaged. At 36, my youngest just entered preschool. At 49, my son was in college and my daughter was just a year away. And my next 'reflection' year is at age 64, ten years from now. Perhaps that will coincide with grandchildren, as my children will be 32 and 34.

As I approached 49, I began thinking about what I wanted to do with the next 15 years of my life. And I realised that games could provide a framework, or perhaps even a philosophy, to live by. Perhaps it was time to start thinking about switching from building my engine to gaining victory points. Perhaps I needed to make my own pivot point.

I've spent a lot of time building my engine. To me that means learning all that I can, reading all that I can, making connections, and trying to understand the world a little bit more each day. And I think that, for me, I earn victory points by passing that knowledge onto others. Building my engine is the input, and victory points are the output. And part of my output is that I believe that the universe is understandable, and that expanding that understanding is, on the whole, beneficial for humanity.

Endnotes

Chapter 1: How to Win at *Rock-Paper-Scissors*

1 Zhijian Wang, Bin Xu and Hai-Jun Zhou, 'Social Cycling and Conditional Responses in the Rock-Paper-Scissors Game' (April 2014) *Scientific Reports*.

2 Richard Cook *et al.*, 'Automatic Imitation in a Strategic Context: Players of Rock-Paper-Scissors imitate opponents' gestures' (July 2011) *Proceedings of the Royal Society B: Biological Sciences*.

3 Antoine Bechara *et al.*, 'Insensitivity to Future Consequences following Damage to Human Prefrontal Cortex' (1994) 50(1–3) *Cognition* 7.

Chapter 3: Game Theory

1 David Hume, *A Treatise of Human Nature* (1738).

2 For a great discussion of the concept of the 'Magic Circle', see Katie Salen Tekinbaş and Eric Zimmerman, *Rules of Play* (MIT Press, 2003).

Chapter 4: Memory, Choice and Perspective

1 Lauren A Leotti and Mauricio R Delgado, 'The Inherent Reward of Choice' (2011) 22(10) *Psychological Science* 1310.

2 Sheena S Iyengar and Mark R Lepper, 'When Choice is Demotivating: Can One Desire Too Much of a Good Thing?' (2001) 79(6) *Journal of Personality and Social Psychology* 995.

3 Paul P Maglio *et al.*, 'Interactive Skill in Scrabble' (1999) *Proceedings of Twenty-first Annual Conference of the Cognitive Science Society*.

Chapter 6: Feeling the Loss

1 Daniel Kahneman, Jack L Knetsch and Richard H Thaler, 'Experimental Tests of the Endowment Effect and the Coase Theorem' (1990) 98(6) *Journal of Political Economy* 1325.

2 Brian Knutson *et al.*, 'Neural Antecedents of the Endowment Effect' (2008) 58(5) *Neuron* 814.

3 Joseph Nunes and Xavier Dreze, 'The Endowed Progress Effect: How Artificial Advancement Increases Effort' (March 2006) 32 *Journal of Consumer Research* 504.

4 For more on endowment and endowed progress, see Jamie Madigan, *The Psychology of Video Games*, https://www.psychologyofgames.com/author/jamie-madigan/ and Chip Heath and Dan Heath, *Switch: How to Change Things When Change is Hard* (Currency, 2010).

5 RE Knox, and JA Inkster, (1968). 'Postdecision dissonance at post time' (1968) 8(4, Pt 1) *Journal of Personality and Social Psychology* 319.

6 Jason Rosenhouse, *The Monty Hall Problem: The Remarkable Story of Math's Most Contentious Brain Teaser* (Oxford University Press, 2009).

Chapter 7: Who's the Best Judge?

1 Daniel Kahneman, *Thinking, Fast and Slow* (Farrar, Straus & Giroux, 2011).

2 Nate Straight, '"Everyone's a Critic!": On Approaching Metacritical Mass', 13 April 2012, *BoardGameGeek*, https://boardgamegeek.com/blogpost/9539/everyones-critic-approaching-metacritical-mass.

Chapter 8: Falling for Patterns

1 Linda Carli, 'Cognitive Reconstruction, Hindsight, and Reactions to Victims and Perpetrators' (1999) 25(8) *Personality and Social Psychology Bulletin*.

2 Baruch Fischhoff, 'Hindsight is not equal to foresight: The effect of outcome knowledge on judgment under uncertainty' (1975) 1(3) *Journal of Experimental Psychology: Human Perception and Performance* 288.

Chapter 10: Games and Entropy

1 Dave Bayer and Persi Diaconis, 'Trailing the Dovetail Shuffle to its Lair' (1992) 2(2) *The Annals of Applied Probability* 294.

2 Sam Assaf, Persi Diaconis, K. Soundararajan, 'A Rule of Thumb for Riffle Shuffling' (2011) 21(3) *The Annals of Applied Probability* 843.

3 For more about autocatalytic sets, check out Stuart Kauffman, *At Home in the Universe: The Search for the Laws of Self-Organization and Complexity* (Oxford University Press, 1996).

Chapter 11: The Evolution of Life

1 Justin Kruger and David Dunning, 'Unskilled and Unaware of It: How Difficulties in Recognizing One's Own Incompetence Lead to Inflated Self-assessments' (1999) 77(6) *Journal of Personality and Social Psychology* 1121.

2 Mark E Glickman and Thomas Doan, 'The US Chess Rating System', 24 April 2017, *US Chess Federation*, http://glicko.net/ratings/rating.system. pdf.

Chapter 12: What's Good about Being 'Good'?

1 Joseph Henrich *et al.*, 'Costly Punishment across Human Societies' (2006) 312(5781) *Science* 1767.

2 One example is Alain Cohn, Ernst Fehr and Michel André Maréchal, 'Business Culture and Dishonesty in the Banking Industry' (2014) 516(7529) *Nature* 86.

3 One example is Nina Mazar and Dan Ariely, 'Dishonesty in Scientific Research' (2015) 125(11) *The Journal of Clinical Investigation* 3993.

Chapter 13: *Tic-Tac-Toe* and Entangled Pairs

1 Allan Goff, 'Quantum Tic-Tac-Toe: A Teaching Metaphor for Superposition in Quantum Mechanics' (2006) 74(11) *American Journal of Physics* 962.

2 Here are two explanations of Bell's Theorem that may help you: http://theworld.com/~reinhold/bellsinequalities.html and http://drchinese.com/David/Bell_Theorem_Easy_Math.htm.

Chapter 14: When Math Doesn't Have All the Answers

1 Alfred Whitehead and Bertrand Russell, *Principia Mathematica* (Cambridge University Press, 1910).

2 Kurt Gödel, 'On Formally Undecidable Propositions of Principia Mathematica and Related Systems' (translated title), (1931) 38(1) *Monatschefte fur Mathematik und Physik 173*. For a great layman's discussion of Gödel's work, I highly recommend Douglas Hofstadter, *Gödel, Escher, Bach: An Eternal Golden Braid* (Basic Books, 1979).

Chapter 15: Goldilocks, Three Bears and a Whole Lot of Noise

1 Ashton Anderson, Jon Kleinberg and Sendhil Mullainathan, 'Assessing Human Error Against a Benchmark of Perfection' (2016) *Proceedings*

of the Twenty-second ACM SIGKDD International Conference on Knowledge Discovery and Data Mining 705.

2 Craig R Fox and Amos Tversky, 'Ambiguity Aversion and Comparative Ignorance' (1995) 110(3) *Quarterly Journal of Economics* 585.

3 Daniel Ellsberg, 'Risk, Ambiguity, and the Savage Axioms' (1961) 75(4) *Quarterly Journal of Economics* 643.

4 Chip Heath and Amos Tversky, 'Preference and Belief: Ambiguity and Competence in Choice under Uncertainty' (1991) 4(1) *Journal of Risk and Uncertainty* 5.

Chapter 17: Making and Breaking the Rules

1 Douglas Hofstadter and Emmanuel Sander, *Surfaces and Essences: Analogy as the Fuel and Fire of Thinking* (Basic Books, 2013).

2 This thought experiment regarding *Incan Gold* and theme originally appeared in an episode of 'GameTek': 'Theme, Mechanics, Experience'. After the segment aired, Stephen Blessing, a Cognitive Psychology professor from the University of Tampa, Florida, contacted me, having decided to investigate it in his next research study. Blessing plans to have participants play different versions of *Incan Gold* to see if their game-playing behaviour changes based on the theme.

3 Gerben Van Kleef *et al.*, 'Breaking the Rules to Rise to Power: How Norm Violators Gain Power in the Eyes of Others' (2011) 2(5) *Social Psychological and Personality Science* 500.

Chapter 18: Community of Players

1 Don Redelmeier and Daniel Kahneman, 'Patients' Memories of Painful Medical Treatments: Real-time and Retrospective Evaluations of Two Minimally Invasive Procedures' (1996) 66(1) *Pain* 3.

Chapter 19: Emergent Properties and Games

1 Rosemarie Nagel, 'Unraveling in Guessing Games: An Experimental Study' (1995) 85(5) *The American Economic Review* 1313.

Epilogue: Pivot Points and Phase Transitions

1 Leonhard Euler, 'Solutio Problematis ad Geometriam Situs Pertinentis' (1741) 8 *Commentarii Academiae Scientiarum Petropolitanae* 128.

Index

Acknowledgments

Over ten years ago, on 25 August 2007, the first 'GameTek' aired on *The Dice Tower* podcast. It has been a wonderful journey. Being part of *The Dice Tower* community and the gaming community at large has been such a positive force in my life.

For this opportunity I am extremely grateful to Joe Steadman, Sam Healey, Eric Summerer and most of all Tom Vasel, for allowing me to play in their playground. *The Dice Tower* will always be my home away from home in the gaming world.

I would also like to thank the team at HarperCollins Australia: Mary Rennie for believing in this book, Shannon Kelly for doing a fantastic job making ten years of writing sound like it was meticulously planned out, and Rachel and Maddy for getting it across the finish line.

My parents gave me my love of games, thirst for learning, and the desire to always try to do a little bit better, for which I am forever grateful. My children, Brian and Sydney, are always up for another game, and always happy to tell me when my ideas need a bit more, shall we say, polish.

And, finally, all my love and appreciation to Susan, who came up with the idea for this book, and has always supported all of my crazy projects.